Resilience, Development and Global Change

Resilience is currently infusing policy debates and public discourses, widely promoted as a normative goal in fields as diverse as the economy, national security, personal development and well-being. This is especially so in responses to climate change, disasters and perceived threats to food, energy and water security. *Resilience, Development and Global Change* critically analyses the multiple meanings and applications of resilience ideas in contemporary society and suggests where, how, why and to what extent resilience might cause us to rethink global change and development, and how a new approach might inform deliberative transformative change.

The book shows how current policies adopting resilience terms and concepts generally promote 'business as usual' rather than radical responses to change. It argues, however, that resilience can help us to understand and respond to the challenges of the contemporary age. These challenges are characterised by high uncertainty, globalised and interconnected systems, increasing disparities and limited choices. Resilience concepts can overturn orthodox approaches to international development that remain dominated by modernisation, aid dependency and a focus on economic growth; and to global environmental change, characterised by technocratic approaches, market environmentalism and commoditisation of ecosystem services. It presents a view of development as transformation, extending current thinking on resilience.

Resilience, Development and Global Change presents a sophisticated, theoretically informed synthesis of resilience thinking across disciplines, including social ecological systems and human development. It proposes a re-visioning of resilience to meet contemporary international development challenges highlighting hitherto neglected areas of resistance, rootedness and resourcefulness. This re-visioning brings novel insights to transform responses to climate change, understandings of poverty dynamics and conceptualisation of social ecological systems. Katrina Brown provides an original perspective for scholars of international development, environmental studies and geography, and introduces new dimensions for those studying broader fields of ecology and society.

Katrina Brown is Professor of Social Sciences at the Environment and Sustainability Institute at the University of Exeter, based in Cornwall in the UK. She has a strong commitment to interdisciplinary analysis of, and innovative approaches to, environmental change and international development.

Resilience, Development and Global Change

Katrina Brown

LONDON AND NEW YORK

First published 2016
by Routledge
2 Park Square, Milton Park, Abingdon, Oxon OX14 4RN

and by Routledge
711 Third Avenue, New York, NY 10017

Routledge is an imprint of the Taylor & Francis Group, an informa business

© 2016 Katrina Brown

British Library Cataloguing in Publication Data
A catalogue record for this book is available from the British Library

Library of Congress Cataloging in Publication Data
A catalog record has been requested for this book

ISBN: 978-0-415-66346-5 (hbk)
ISBN: 978-0-415-66347-2 (pbk)
ISBN: 978-0-203-49809-5 (ebk)

Typeset in Sabon
by Saxon Graphics Ltd, Derby

To James, who has taught me the most important things about resilience, with love.

Contents

Figures

Tables

Boxes

Preface

At the Resilience 2014 conference in Montpellier, France, resilience was described as a new paradigm for development. I do not test this claim here, and I do not want to present resilience as a new paradigm, but I want to explore whether and how ideas from different disciplinary perspectives about resilience can inform international development.

I believe that the assertion about resilience and development, the drive for a new paradigm, stems from gaps in current development science and practice, where development theory and knowledge really do not explain what has happened, and don't give adequate means of integrating sudden shocks, systemic change and uncertainty – be that global financial crisis, environmental change or 'crisis states'. In other words, we do not have intellectual tools to understand how these affect development. A major reason why we need a new approach to understanding development is climate change, which brings challenges so profound and so wide-reaching that the only thing we can expect of the future is that it will be nothing like the present or the past. Climate change threatens current notions and practice of development and demands a new approach to well-being, progress, justice and environment. I see both scholarly and policy imperatives for a new set of ideas about development in this age of uncertainty and recurrent crises. In many ways, resilience itself has entered the policy lexicon before it has really entered mainstream development studies.

My aim is to apply a broadly defined set of ideas around resilience to international development in the face of global change. The book is aimed primarily at scholars, particularly researchers of environmental change and international development; and practitioners, including those who work in large organisations and development agencies devising strategic approaches to addressing climate change and other major disruptions, and those in NGOs tackling poverty and responding to disasters. I hope the book will be of interest to anyone who works in the broad area of sustainable development and I present my proposals as a new way of thinking about sustainability and joining up related fields – environmental change, international development, systems thinking. My hopes for the book are that it inspires integrated thinking and challenges some assumptions about the relationship between change and development, about winners and losers, and about human agency in the face of profound, rapid and irreversible changes that we collectively face.

Acknowledgements

This work has had a long gestation and a protracted delivery. I have benefitted in many, many ways from the support, inspiration, collaboration and help of many people.

I have been extremely fortunate to work with many very talented and accomplished scholars, and I have learned much from them. They include in no small part my postgraduate and research students, and co-researchers on a number of projects. Each of these clever people has been a joy to work with. Those whose work has directly informed this book include, Neil Adger, Jacopo Baggio, Nick Brooks, Matthew Bunce, Tom Chaigneau, Esteve Corbera, Tim Daw, Siri Eriksen, Lucy Faulkner, Roger Few, Natasha Grist, Denis Hellebrandt, Tom James, Sandrine Lapuyade, Don Nelson, Sergio Rosendo, Tara Quinn, Lucy Szaboova, Emma Tompkins, and Liz Westaway.

This research was initially funded by a Professorial fellowship from the Economic and Social Research Council (ESRC Grant number RES-051-27-0263 *Resilient development in social ecological systems*) and I very gratefully acknowledge the support of ESRC. The book also draws on other research, notably that funded by Leverhulme Trust (*Coastal resilience to climate change in eastern Africa*), Tyndall Centre, and latterly the AXA Research Fund (*You, me and our resilience*). At the University of East Anglia (UEA) I benefitted from a uniquely supportive, collegiate and stimulating intellectual environment that was deeply committed to interdisciplinarity and engagement, and gave real scope for innovation. I am indebted to colleagues in DEV and DevCo, and in the Tyndall Centre. At Exeter University I've been very generously supported through the Environment and Sustainability Institute (ESI) and my colleagues there. In particular research leave enabled me to complete the manuscript. I was very kindly hosted by Commonwealth Scientific and Industrial Research Organisation and enjoyed many stimulating discussions with scientists here and at the Australian Research Council Centre for Excellence in Coral Reefs at James Cook University in Australia. My special thanks are due to Nadine Marshall and her family who facilitated and very generously hosted us during our stay in Townsville.

One of the greatest gifts of my career has been that I've been able to work with collaborators around the world. Again, many people have inspired and helped me, and I've enjoyed stimulating conversations and collaborations with too many colleagues to list all of them here. They include Erin Bohensky, François Bousquet, Josh Cinner, Beatrice Crona, Christo Fabricius, Bronwyn Hayward, Christina Hicks, Tim Lynam, Nadine Marshall, Paul Marshall, Karen O'Brien, Charles Redman, Helen Ross, Samantha Stone-Jovicich and Brian Walker.

Laura Middleton has been a wonderfully supportive and effective helper, organising references, diagrams and copyrights. I am very grateful to have had her alongside me on this journey.

Finally my family and especially Neil and James have put up with my long hours and have buoyed me through the writing process.

Any errors are solely the responsibility of the author.

In Chapter 7, Figure 7.1 'Stop calling me resilient': Every effort has been made to contact rights holders.

Please make the Publisher aware of any omissions so this can be corrected in future editions of the work.

Introduction

My motivation to write this book stems from my engagement with 'resilience thinking' primarily from the perspective of environmental change, and finding some of its fundamental premises really exciting, but realising that few of these had been widely applied in international development. I understand international development as the process of managing and shaping change to enhance human welfare and well-being and to overcome poverty. Resilience struck me as being of fundamental importance to conceptualising and effecting sustainable development. Resilience thinking – encompassing ideas from complex adaptive systems – can bring four important dimensions to the understanding of sustainability and formulation of sustainable development.

Resilience concepts bring recognition first that uncertainty is part of how 'systems' work and that we should expect the unexpected; second, that systems are inherently dynamic and there are multiple links and feedbacks between processes and changes; third, that there are important temporal, societal and spatial cross-scale interactions; and fourth, that multiple stressors and drivers act on systems and interact, sometimes with synergistic results. For example, climate change interacts with other stressors such as food or fuel price change to produce specific impacts on specific people. Resilience can help us to better understand these linkages and dynamics, and the trade-offs between different courses of action. Importantly current policy making rarely takes a dynamic view; generally speaking and across the globe, it remains sectoral, linear and short-term. The case made here is that resilience thinking can potentially enhance not only our scientific understanding of social and ecological change processes, but also our policy responses to enhance well-being and life opportunities, particularly of poor people.

So the first part of this book examines the resilience phenomenon in development; it reviews discourses of resilience in development and climate change policy literature and it looks at the everyday experiences and narratives of resilience. It then examines key resilience concepts and how they apply to development, including adaptive capacity and adaptation, poverty and social ecological traps, and transformation.

I present resilience as multi-dimensional, and as pluralistic in its meanings and interpretations. This recognises three key dimensions of resilience:

- bouncing back; regaining stability after a shock – a focus of disaster and humanitarian aid;
- adapting to variability and uncertainty – the focus of much of the climate change and development field;
- positive transformation, requiring structural change, which is currently an under-researched and poorly evidenced aspect of resilience in development.

Resilience is thus broadly defined as the ability to successfully deal with change, and it is a characteristic that can be applied to individuals, communities, states, ecosystems or linked social ecological systems, tightly coupled systems of people and environment.

Resilience is a concept that scientists feel strongly about. Importantly social scientists criticise resilience applied to social ecological systems for neglecting politics, power and agency, and for being naive and not suitably nuanced in its understanding and appreciation of social dynamics. One of the aims here is to develop thinking around resilience that fully integrates these aspects, and thus sets resilience to work within some of the core concerns of development studies – for example, social justice, equity and well-being. Resilience has been hailed as the latest 'buzzword' and has risen to prominence in scientific, policy and public media. I examine the different meanings of resilience, and why and how resilience has become so popular and so widely adopted. The framings of resilience, in terms of sustainability agendas and in response to heightened concerns about risk, uncertainty and insecurity are explored in this book. Chapter 1 sets out and explains how and why a political ecology approach brings new insights and perspectives to understanding, and adds to current knowledge and debates in the field of development and change.

Importantly, we need to understand how resilience ideas are currently being applied in the fields of international development and global change. Chapter 2 examines the policy discourses that are emerging from various international, UN and non-governmental bodies. It discusses how resilience is used – and certainly was used until very recently – to justify and bolster the 'status quo'. This appears quite counter to resilience as a dynamic property of systems that supports beneficial change and development.

Resilience is a concept that is used in diverse scientific fields – including (but not exclusively) engineering, ecology, child developmental psychology and, increasingly, in areas of health and disasters, and social ecological systems. Chapter 3 reviews the evolution in thinking across these fields, and identifies some of the similarities in terms of key concepts, such as adaptation, capacity, feedbacks and dynamics. Whilst there are some important shared core concepts, there are also some critical differences. Resilience may be an important boundary object or boundary concept – in other words, rather like sustainability, it is able to form a bridge across areas of science and policy. However, the evidence is limited to suggest this, and many ideas are 'lost in translation' between science and policy.

But how is resilience 'lived', especially by poor people in poor countries, who face growing uncertainty, rapid change and increasing and novel vulnerabilities? Chapter 4 applies a political ecology lens to understanding the dynamics of resilience and vulnerability. It employs a series of vignettes that highlight the critical issues that emerge from empirical application of resilience thinking. It presents analysis of peoples' perceptions, narratives and mental models of resilience; what resilience means for them and how it is enacted. Resilience is revealed to be highly socially differentiated and often contested. This discussion grounds resilience within the more conventional concerns of international development, around power, justice and equity, poverty and well-being. It documents examples of the 'everyday forms of resilience', which is used as the basis for re-visioning resilience for development.

Resilience brings new insights for science and policy, and by applying dynamic systems thinking – examining cross-scale interactions between multiple stressors – can bring fresh perspectives on vulnerability, adaptation and development in the context of climate change. Applying these concepts highlights some current responses as maladaptive. This is illustrated by how current approaches to addressing climate change may enhance vulnerability, particularly of the poor and the marginalised. Chapter 5 explains how resilience analysis can help in developing policies that enhance adaptive capacities and overcome these maladaptations, for more sustainable adaptation to climate change.

My vision of resilience applied to international development in the era of global change revolves around ideas about positive transformation. Chapter 6 discusses how concepts such as poverty traps and rigidity traps can help to understand what keeps people in poverty and identify the possible routes out of poverty. It reviews some of the emerging science around transformations and how this might contribute to a new agenda for development. This runs counter to some of the recent applications of resilience evidenced in earlier chapters. It presents a radical interpretation and application of resilience ideas.

The final chapter, Chapter 7, elaborates the notion of everyday forms of resilience, which brings together individual agency and systems perspectives. This vision of resilience has three core components; resistance, rootedness, and resourcefulness. Resistance puts concerns for politics and power at heart of resilience. It concerns how new spaces for change can be opened up and how positive transformation might be shaped and mobilised. Rootedness acknowledges the situated nature of resilience, and the importance of culture and place – not only as physical environment and context, but also as identity and attachment. Rootedness de-centres resilience strategies and locates them, whilst also working at and across multiple scales. Resourcefulness considers the resources available, how they can be accessed and used in response to change. It concerns capacities, knowledges, innovation and learning. But these three 'Rs' are not enough for a new development agenda. The context in which a re-visioned resilience is used and how it can inform and support decisions and planning for the future are critically important. A re-visioning of resilience also potentially re-configures conventional understanding of human-environment relations in a social ecological system, and contributes to re-positioning development in the era of global change.

1   Resilience now

Resilience has recently been described as a theory of change, a new development paradigm, a defining metaphor for our era, and a buzzword. Clearly the term and the concepts around it have significant resonance and the traction for current thinking and policy on global change, development and environment. The concepts of resilience, development and transformation are the central subjects of this book. This chapter sets out the key arguments and justification for the book, and discusses some of the contemporary framings of resilience. This situates resilience very firmly in a transdisciplinary arena addressing the urgent need for a new set of guiding principles and concepts to inform international development in the age of perceived rapid and large-scale global changes. It presents a number of definitions of resilience and related concepts, and some of the main critiques of resilience science and the policy prescriptions that flow from it. It shows why a political ecology approach – which examines the relationships between political, economic and social factors with environmental issues and changes – gives rise to a constructive engagement with, and opens new avenues for, resilience.

Why resilience, why now?

Resilience has become increasingly prevalent in both scientific and policy realms and in public discourse. Resilience, it seems, is everywhere, highly prominent in both scientific and popular debates. In the wake of any sudden event or disaster, there are inevitably calls for increased resilience or narratives about how resilient people and communities, ecosystems or cities – even the 'economy' – are in the face of a shock or calamity. Table 1.1 shows recent examples of these proclamations in the media about resilience in different contexts from various arenas of public life. This rise in the resilience has taken place particularly in the last decade and especially since the global financial crisis in 2007–8. Recent analyses show continued and steady rise in the popularity of the term, for example as a search term on Google, and in terms of published scientific articles (Baggio et al. 2015; Xu & Marinova 2013).

Table 1.1 Resilience in public discourse

Event	Quote
Pakistan floods, 2011	'The People of Pakistan have shown remarkable strength and resilience throughout the disaster, supporting each other to overcome extraordinary adversity.' Rauf Engin Soysal, Special Envoy of the United Nations Secretary-General for Assistance to Pakistan
Financial crisis, 2011	'We can't hope to prevent financial crises from happening, but we can build institutions that help to ensure that our financial system is more resilient in the future.' Mervyn King, former Governor of the Bank of England, on preventing financial crises
World Economic Forum, 2013	'Dynamism in our hyper-connected world requires increasing our resilience to the many global risks that loom before us.' Klaus Schwab, Founder and Executive Chairman, World Economic Forum
Hurricane Sandy, 2013	'... even as we continue to work with those communities today, it is valuable to assess the lessons learned from this natural disaster so that we can rebuild stronger, more resilient communities that are better prepared for any future extreme weather.' Maria van der Hoeven, Executive Director of the International Energy Agency, US Department of Energy
US National Disaster Resilience Competition, 2014	'President Announces $1 Billion Climate "Resilience" Fund to Help Communities Prepare for Natural Disasters.' *International Business Times*

Source: Author's own

Resilience is being promoted not just in relation to how people can respond to catastrophic or extreme events – 'shocks' to the system – but also to describe proactive adaptation and anticipatory action. In this way, resilience is understood not only as a response to change, but also as a strategy for building the capacity to deal with and to shape change. In this book I explore some of these different applications of resilience terminology and ideas, and focus specifically on international development, mainly – though not exclusively – in developing countries, and chiefly in the context of environmental and climate change. Development is understood here to mean the process of cultural, demographic, economic, political and social change, with a particular focus on the reduction or elimination of poverty in poor countries.

The discussion covers how resilience is increasingly applied in both scientific and public discourse. In this chapter I discuss how resilience has many different meanings, interpretations and applications. But as the applications and prominence of resilience increase, so does the need to understand its diverse framings, narratives and discourses (Leach 2008). The aims of the book are to make sense of these diverse meanings and synthesise key issues, and to distil the novel aspects of resilience that could inform a new approach to understanding, managing and shaping change and development. This chapter introduces resilience and its different meanings. It explains the analytical lens applied to resilience and change, setting out a broadly political ecology approach. This acknowledges the diverse meanings and different types of knowledges that construct them, as well as the contestations and claims surrounding them. It recognises socially constructed discourses, as well as more realist meanings, and helps to distinguish and evaluate both the normative and analytical dimensions, and to give weight to the empirical and policy applications, and the everyday lived experiences of resilience.

Defining resilience

Many studies have reviewed the different meanings of resilience across fields. Box 1.1 shows three definitions from three distinct scientific fields. This is a highly selective set of resilience definitions. Of course, there are many other definitions, and the term is used in other fields and with different emphases and meanings. Indeed, a number of publications compile and comment on long lists of definitions (for example Martin-Breen & Anderies 2011; Bahadur et al. 2010). Chapter 3 discusses how the concept of resilience

---

Box 1.1 Resilience definitions from different scientific fields

Resilience is...

'The ability to absorb disturbances, to be changed and then to re-organise and still have the same identity (retain the same basic structure and ways of functioning).'

'In the context of exposure to significant adversity, resilience is both the capacity of individuals to navigate their way to the psychological, social, cultural and physical resources that sustain their well-being, and their capacity individually and collectively to negotiate for these resources to be provided in culturally meaningful ways.'

'A multi-dimensional construct . . . the capacity of individuals, families, communities, systems and institutions to respond, withstand and/or judiciously engage with *catastrophic* events and experiences; actively making meaning without fundamental loss of identity.'

Source: Author

has evolved across these different fields. But the three definitions here encapsulate a range of the broad interpretations and popularly understood meanings. The first is from the Resilience Alliance,[1] and focuses on resilience in social ecological systems, which emphasises the integrated system of people and environment, with understanding of resilience derived from ecology but encompassing broader related fields of environmental change and natural resource management. The second is from Michael Ungar, Professor of Social Work who co-directs the Resilience Research Centre.[2] The third definition comes from the broader perspective of public health, articulated in an editorial from the journal *African Health Services*.[3] These three perspectives, encompassing social ecological systems, human development and applied fields, inform the view of resilience in this book.

These definitions have important similarities. They each identify resilience as *capacity* – of an individual, community or a system. In addition resilience is both a *process* and an *outcome*. It involves not only some notion of staying the same – in terms of maintaining identity or functioning – but also undergoing change, actively engaging in change or adapting to change. One of the most straightforward definitions is from Ann Masten, a leading psychologist working on resilience. She defines resilience as 'the process of, capacity for, or outcome of successful adaptation despite challenging or threatening circumstances' (Masten et al. 1990: 425). The significance for my own research is that resilience is a property of individuals, households, communities and social ecological systems. Secondly, resilience is capacity, process and outcome.

These issues are discussed in more detail in Chapter 3, but they reveal some of the multiple dimensions and nuances which make resilience so interesting, but also potentially misunderstood and misapplied. As the application of the term expands, across scientific fields and into policy and public discourse, so its meanings get stretched. There is plenty of room for contestation and confusion. For example, there is a critical distinction, and very often confusion, between normative applications that are generally prescriptive and assume resilience is always desirable, and theoretical or descriptive approaches where resilience is neither inherently good nor bad. Importantly there are distinctions in approaches and emphases on whether resilience is about staying the same or changing in response to disturbance; and whether resilience is a process or an outcome of exposure to trauma. Clearly some of these differences might be quite fundamental and might lead not only to different and distinct interpretations, but also suggest or result in different courses of action.

Resilience ideas have emerged in science in the last 40 years, yet have been popular in many areas of policy primarily in the past decade. Table 1.2 shows some of these meanings and applications in different fields and the different concepts that are emphasised and used in them. It also introduces

Table 1.2 Applications and core resilience concepts in different fields

Field	Applications	Concepts
International relations	Understanding military and terrorist threats	Security Critical infrastructure
Social ecological systems	Managing complex systems in times of change Informing adaptive management strategies	Adaptive cycle Adaptive capacity Transformations Linking social and ecological dimensions of resilience
Disasters and disaster risk reduction	Minimising risk and support recovery	Vulnerability Community resilience
Climate change	Adapting to and minimising impacts of climate change	Adaptation Adaptive capacity Climate resilience
Human development	Coping and thriving in times of adversity Individual responses to crises Poverty traps	Individual resilience Human well-being Capacity Agency
Organisational science and social innovation	Managing change	Social learning
Planning	Urban and regional planning	Urban resilience

Source: Brown 2014: 108

key related concepts and ideas (also see Glossary). In many respects, the concept is still evolving and is developing many hybrid meanings, many applications, and context-specific interpretations. Commentators and analysts debate whether resilience is a boundary object or a buzzword, its analytical depth, and the extent to which shared meanings are apparent. Importantly reading resilience literature across broad fields, including psychology, public health and ecology, also identifies important commonalities, so it is possible to synthesise a resilience approach and the elements it encapsulates. This is summarised in Box 1.2 and constitutes the understanding and approach adopted here.

This book explores how this approach is applied across many different areas of policy and practice in international development (Chapter 2) and in different fields of science (Chapter 3). I take an inclusive approach to understanding resilience, so that resilience encompasses the ability to withstand, to bounce back from, and to emerge more strongly from shocks and change. The Rockefeller Foundation, which uses resilience as a central theme in its work, and especially in its 100 Resilient Cities project, has a useful definition that takes an inclusive view. But this is a normative stance,

---

Box 1.2 What does a resilience approach highlight?

- Expect change, manage for change – leads to a prescriptive focus on adaptive management.
- Expect the unexpected – uncertainty and surprise are features of systems.
- Recognise different types of change; slow and fast variables and the interactions between them.
- Crises may provide windows of opportunity – the chance to move to a new regime which may be either better or worse than the existing one.
- Thresholds are a feature of change and most change is not uniform, regular or predictable – thresholds are ecological and social, and may be manifest as 'tipping points'.
- Multi- and cross-scale issues are important – understanding the interplay and links and interactions is a challenge and has led to examination of polycentric institutions and the concept of 'panarchy'.
- Interactions with other stressors – climate change, livelihoods, health, markets, migration and settlement – are recognised by the systems lens – but there may be both general and specific forms of resilience.
- Resilience can be 'good' or 'bad' – it can lead to rigidity or 'traps'.

Source: Author's own

---

showing how resilience is applied and promoted, rather than a scientific definition:

> We define resilience as the capacity of individuals, communities, and systems to survive, adapt, and grow in the face of stress and shocks, and even transform when conditions require it. Building resilience is about making people, communities, and systems better prepared to withstand catastrophic events – both natural and manmade – and able to bounce back more quickly and emerge stronger from these shocks and stresses.
> (Rockefeller Foundation 2013)

This inclusive view encompasses at least three dimensions of resilience. First is the ability to resist, cope and bounce back in the face of disturbance. This aspect is conventionally the focus, for example, of a disaster risk reduction approach. It is a view often reflected in popular understandings and conservative approaches to resilience, which the next two chapters explore in more detail. Second is the capacity to adapt to change and uncertainty, which has been a major focus of climate change adaptation approaches in development. These are analysed further in Chapter 5. Third is the capacity for transformation, to radically change in order to take advantage of new opportunities and new possibilities – to positively develop and to thrive in the face of change and uncertainty. The first two aspects of resilience – coping and adapting – have been reasonably well studied and observed empirically. They already have been embraced and have strong purchase in international development. The third aspect, a more radical, transformative dimension, has far less empirical study or evidence and has yet to be transferred to meaningful and feasible policy. It is a dimension of resilience around which literature is starting to emerge, and where exciting and innovative opportunities for rethinking development may lie. The central premises or principles of resilience – the points outlined in Box 1.2 – are helpful in reformulating development in an age of rapid global change and widespread uncertainties. Although already applied in this field to an extent, resilience concepts are still under-developed, particularly in terms of the transformative outcomes of resilience. This remains a relatively new dimension in resilience studies and to date has been overlooked in favour of focus on the role of resilience in coping with and adapting to shocks and other changes. But the transformational or transformative dimensions have significant implications for development theory and practice.

As long ago as 1998, Lélé observed that 'resilience is turning out to be a resilient concept' (p249), but more recently there have been claims that resilience is a new paradigm for development. Whilst this book does not set out to test this claim, it explores how resilience ideas are being taken up by development scholars and in development practice, and proposes how resilience ideas might inform and support a new development agenda. Lawrence Haddad questions whether resilience is a 'mobilising metaphor' for twenty-first-century development (Haddad 2014) and observes that much of our thinking and theorising about development developed in the last half of the twentieth century, a world very different from today. So perhaps a concept which encompasses ideas about systemic change, shocks and interconnections can prompt a radical shift in thinking. Béné et al. (2014) remark that not only is resilience gaining critical mass in academic arenas, it is also becoming prominent across the whole development agenda. But Béné and colleagues are adamant that resilience is not a pro-poor concept and that therefore it has serious limitations for development. In a similar vein, Sharachchandra Lélé, who has written extensively about sustainable development, identifies shortcomings in resilience particularly related to intra-generational equity, and makes an important point that 'no amount of concern for long-term resilience of the human ecosystem can by itself ensure a fair environmentalism or a just development' (1998: 253). Later sections of this book dissect and discuss these claims and criticisms in greater detail. In particular, Chapter 2 examines how international development agencies are using resilience terminology in strategic policy documents and how they are applying resilience concepts in practice.

The following sections explain the analytical approach to resilience taken here, and how and why this adds to current perspectives. It outlines dominant framings of resilience, and questions how resilience might be re-framed to make it speak more directly, lucidly and helpfully to concerns for development in the twenty-first century.

A political ecology approach

The discussion and definitions outlined already point to some of the areas and issues surrounding resilience where there are different meanings and interpretations, and ambiguities. In reality, many ideas in resilience are contested. In particular, there are tensions between positivist accounts of resilience on the one hand and the politics of knowledge in the construction of resilience theories and models on the other. There are also criticisms of how poorly resilience ideas from systems perspectives and the natural sciences (such as ecology) translate into social contexts, and whether they can incorporate or add to understanding the dynamics of social systems. These are discussed in a recent report for *Progress in Human Geography* (Brown 2014), and summarised in Box 1.3.

Box 1.3 Key criticisms of resilience thinking

- Resilience thinking ignores social dynamics and rarely asks 'resilience for whom'?
- Resilience approaches focus on maintaining the status quo, on reinforcing existing power relationships and structure and not addressing root causes.
- A resilience lens focuses on short-term stability rather than long-term sustainability.
- A systems approach foregrounds exogenous drivers of change rather than internal or endogenous drivers.
- Resilience thinking fails to account for power and politics.
- Who defines the 'desired state'; whose needs count?
- A resilience approach privileges the technical over the social.
- Resilience is vague and normative.
- Resilience thinking is contradictory and full of inconsistencies.
- The field was developed by a small network of scholars – the Resilience Alliance who have 'discursive dominance'.

Source: Author's own

Issues of power and social difference are central to these criticisms; many criticisms of resilience theory suggest that issues of power, or even social relations, are largely absent from resilience theory (Leach 2008). Hornborg (2009) and others have suggested that policy prescriptions on resilience fail to recognise asymmetries of power between actors and are thus relatively weak and benign towards the status quo (Fabinyi et al. 2014). But there are diverse views within resilience thinking: Walker et al. (2009) for example, argue for a shakeup of global institutions such as World Bank, IMF and UN agencies, and greater co-ordination between them to help construct and maintain a global-scale social contract. Others suggest that resilience requires moving away from globalised governance structures towards greater local autonomy and diversity (O'Brien et al. 2009). Hatt (2013) adds to the critical social science views on the extension of resilience thinking to social systems. His paper speaks to common criticisms about the lack of attention to issues of power in resilience thinking and makes an important point about the anachronism of resilience approaches which adopt a more dynamic, adaptive and multi-equilibria view of ecological aspects, but then apply a simplistic, static and deterministic view of social aspects. Hatt argues for the need for a more dynamic view of social relations and how they shape, mediate and define social ecological systems. From his critical realist perspective, Hatt identifies a crucial role for social science analysis of resilience of social ecological systems and environmental change. The political ecology perspective developed here contributes to this social science analysis. Thus the context and starting point for my analysis is a broad set of views, diverse and evolving meanings, and a rapidly expanding scientific field of resilience.

Political ecology examines the relationships between political, economic and social factors with environmental issues and changes. It incorporates an analysis of power and politics, knowledge and knowing, rights, access and justice to understandings and applications of resilience. This adds to resilience scholarship in a number of key ways and provides new insights into how resilience can be applied to discussions of global change and international development by introducing some of these issues – central to the field of development studies – but hitherto underrepresented in discussions about resilience. Political ecology provides an agenda to analyse the multiple meanings of resilience and to take a politicised and systems perspective on environment, development and global change.

To date there have been limited attempts by political ecology scholars to engage meaningfully with resilience thinking. Generally, political ecology has tended to provide a critique of resilience rather a constructive engagement – to the extent that a debate on Political Ecology and Resilience at Uppsala University in Sweden University in 2012 (Peterson and Hornborg 2012) seemingly set the two approaches in direct opposition. The discussion between the protagonists – Garry Peterson from Stockholm Resilience Centre and Alf Hornborg from Lund University – was often more about *how* the science itself is constructed, rather than about whether and how the approaches provide analytical tools to analyse environment and social relations and change, or contribute towards sustainable development. This discussion exemplifies some of the entrenched critical views – positing 'resilience theory' and political ecology as alternative ways of analysing sustainability, suggesting incompatible epistemologies and methodologies. There are even claims that resilience scholars are colonising political ecology and that by its failure to challenge the status quo and to engage in research on controversial issues, resilience research itself becomes a means of perpetuating the status quo.

As I discuss in more detail in Chapter 2, resilience ideas and concepts are indeed evoked to keep 'business as usual', but I would also claim, and map out here, how resilience theory can be used in subversive ways. This interesting and challenging space at the intersection of the different disciplinary and scientific approaches is particularly rich and is where I position this work. Resilience concepts present potentially powerful and innovative ways of understanding change processes and for re-assessing the role of risk and uncertainty. Political ecology reminds us of the importance of power, winners and losers, plural perspectives and narratives, and to always question the framings and values behind knowledge claims and science. In this way I think the two approaches (neither of which I view as a 'theory') are complementary. They each take a systems approach (although often conceptualised very differently), and both are concerned with 'analysing and responding to environmental challenges and developing a systemic understanding of causes and limitations', according to Lawhon and Murphy (2011: 365), even if they have quite different emphases and antecedents.

Some of the earliest writing in resilience and social ecological systems literature[4] to take a political ecology approach is Peterson's own paper from 2000, and work by Lebel and colleagues emphasising power and knowledge asymmetries (2006), and Fabricius et al. (2007) highlighting the need to understand winners and losers in resilience approaches to resource management. More recently political ecologists have written more extensively, and more critically, about resilience (Beymer-Farris et al. 2012). Since 2012 there has been a stream of work from different fields within social sciences that have dissected resilience (see those reviewed in Brown 2014; and various journal special issues). Fabinyi and colleagues (Fabinyi et al. 2014) usefully show how political ecology can add to resilience scholarship in directly addressing power and diversity, and Robards et al. (2011) put politics at the centre of their discussion of resilience of ecosystem services. So the interfaces between these approaches are opening up and increasing.

Peterson's 2000 paper in the journal *Ecological Economics* is one of the earliest papers to explicitly bring together political ecology and resilience. This paper critiques political ecology for failing to include a thorough consideration of ecology. Peterson posits applying resilience concepts – such as the adaptive cycle and cross-scale dynamics – to strengthen political ecology analysis of human ecological interactions and dynamics. He develops an analysis of management of the Columbia River fishery. His Figure 6, reproduced in this chapter as Figure 1.1, synthesises a resilience view of panarchy (discussed further in Chapter 3, see Glossary) with Stephen Lukes' conceptualisation of power. Peterson uses this to analyse the different scales – social and temporal – at which power operates: thus overt power operates in the here and now, covert power at larger institutional scales, and structural power at the slowest and broadest scales. In this analysis, Peterson argues that Lukes' (1973) three dimensions of power can be thought of as operating at three different scales and that power will follow an adaptive cycle, as shown in Figure 1.1. Pritchard and Sanderson (2002) also use Lukes' three dimensions of power, linking them to fast and slow variables, and to how they determine which models of management of social ecological systems are considered and how scientific and political agendas are controlled and manipulated. In the example of the Columbia River Basin that Peterson uses, overt power operates for brief periods in a specific location – he cites the arrest of Native American 'illegal' fishers by fisheries police. Covert power controls whether issues are discussed or addressed by institutions; in the Columbia River Basin example, this was exercised in how dam building authorities ignored native treaty rights and destroyed native fishing grounds. Structural power operates at the slowest and broadest scales by manipulating cultural norms and values and by the framing of issues. In Peterson's case this concerns how the questioning of whether dams were necessary, and the possibility of removal of dams, was effectively kept off the agenda.

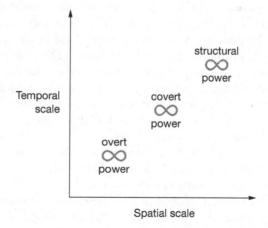

Figure 1.1 Three dimensions and scales at which power operates
Source: Peterson 2000

How then do I characterise my approach as 'political ecology'? There are at least three prevailing views of political ecology, as explained very succinctly in Tim Forsyth's book *Critical political ecology* (2003). These are: Piers Blaikie's political economy of natural resource management; an ecological approach to politics, delineating a systems approach; and the post-structuralist approaches as promoted by Richard Peet and Micheal Watts (1996). The approach here borrows from each of these, and develops an analysis that combines elements from them all. Thus I adopt an archetypal political ecology analytical lens by examining the views and interests of different stakeholders, and identify winners and losers associated with different strategies, particularly in Chapter 4. I analyse the knowledge and power dimensions of resilience claims, and the framings, discourses and narratives surrounding resilience portrayed and promoted by different actors in this and the next three chapters. And I examine how this results in different policy outcomes.

Peet and Watts' (1996) synthesis of political ecology sets out a research agenda for political ecology that provides an excellent platform from which to develop an analysis of resilience, global change and international development. They identify five elements – or what they term 'loosely configured areas of scholarship' (1996: 9) – which have extended the frontiers of political ecology, and I use these to shape my analysis and develop my arguments around resilience. I discuss these in turn below: political economy; politics; environmentalism and environmental movements; discourses and knowledges; environmental history; and ecology.

Peet and Watts identify a political economy component in political ecology that seeks to make explicit and analyse the causal connections between the dynamics of economic growth and development, and specific environmental outcomes. This sets a context for understanding how different policies and framings lead to impacts for different sectors of society and for the environment. It also sees a role for the state, and indeed perhaps supranational powers, in promoting particular forms of growth and development, and thus focuses on structural aspects of society-nature relations. In analysing policy discourses of resilience, the next chapter shows how particular meanings of resilience are linked to assumptions about growth and development, and ultimately to a capitalist economic paradigm. Later chapters discuss how a more radical resilience approach, shaped by this political ecology analysis, might challenge some of the assumptions about growth, and green growth and development, and suggest some quite different priorities for international development.

The second area that Peet and Watts claim to be under-developed in political ecology is that of politics itself. Much empirical work in political ecology has identified issues of power, resistance, and access as central to determining society-nature relations, focusing on land degradation, forestry and conservation in particular. Work on the ecology of the poor, on social difference, and on how different social actors are able to make claims and negotiate access to resources, are core to political ecology analysis. However, the extent to which these works develop insights beyond the empirical and to inform wider theoretical and conceptual dimensions of politics has been questioned. Most importantly the lack of serious consideration of politics is one of the criticisms made most often of resilience approaches. The empirical studies discussed in Chapter 4, which elaborate 'everyday forms of resilience' by analysing experiential resilience, highlight and foreground these dimensions.

Understanding environmentalisms and environmental movements, including diverse knowledges and practices, and their role in providing alternative ways of understanding social ecological systems and as the basis of sustainable livelihoods, is also a central tenet of a political ecology approach. I provide a perspective on experiential resilience and explore how people are able to express, articulate and enact their diverse perspectives on environmental change. Part of this work has involved looking at mental models of social ecological systems, and the dynamics and causes of change within them, and how people view their own agency and self-efficacy. This is examined particularly in relation to adaptive capacity and transformative change and development. In this analysis I highlight some of the different claims about resilience and some of the winners and losers of resilience approaches, and relate this to power and politics. This approach links resilience to resistance and shows how these two are mutually reinforcing and may be used in novel ways to support alternative strategies for environmental management.

This plurality of perceptions and definitions of change and resilience is further explored through discourse analysis, the fourth theme which Peet and Watts highlight. Discursive approaches have been significant in political ecology in presenting critical studies of science and regulatory knowledge; and examining how particular knowledges are privileged and become institutionalised. Particularly important is the globalisation of environmental discourses and the formation of a set of assumptions and frames around global scale environmental change (Taylor & Buttel 1992) with their associated narratives, leading to technocratic and managerialist 'solutions' to environmental 'problems' (Adger et al. 2001). In Chapter 2 I analyse discourses around resilience and how they support a set of solutions that are quite counter to scientific understandings of the dynamics and multi-dimensional and complex nature of environmental change.

Historical perspectives within political ecology bring new insights into understanding change; giving historical depth and an opportunity to chart environmental relations within particular trajectories of development. They have enabled a dynamic understanding which moves beyond a snapshot of recent change and which brings oral and other histories to the fore. By bringing together different historical evidence, and in highlighting the long-term co-evolution of people and environment in place, they often provide counter narratives to the linear and simplistic accounts of environmental change. I use longitudinal studies in Chapter 4 to look at the social relations and differences that underpin and play out in environmental change. Furthermore this historical perspective leads political ecology to engage much more directly with ecology and to interrogate the meanings and assumptions underpinning notions about ecosystems, stability and sustainability. Political ecology authors have been influential in advancing 'new ecology' during the 1990s based on multiple equilibria rather than single-equilibria approaches, which resonates with the challenges of stability models within ecology which resilience concepts represented, where the possibility of multiple states of a system was suggested. This led to challenges of many assumptions about carrying capacity, over-harvesting and environmental degradation. It also promoted a more nuanced understanding of the role of human interventions in ecosystems – which included a much more reflexive view of how active management relates to changes in ecosystems. This is a central theme running throughout this book. But I – along with other writers (e.g. Hatt 2013) – contend that resilience approaches within 'new ecology', whilst developing a dynamic and system-based approach to understanding change in ecological aspects of social ecological systems, have not to date applied the equivalent frameworks to understand complexity and dynamics of social dimensions of these systems.

Political ecology, in acknowledging cross-scale linkages and dynamic patterns of causation seems in many respects complimentary to resilience studies. A political ecology of resilience adds to current knowledge by providing perspectives on power, knowledge and discourse applied to global change. This political ecology approach starts by reflecting on how resilience is framed and how this sets the scene for its predominance in policy, and salience in scientific arenas around environment and development.

Framing resilience

This section discusses some of the different framings of resilience as a prelude to the analysis presented in the three chapters that follow. Framing refers to the social construction of a phenomenon by media, political or social movements, political leaders or other actors or organisations. It defines the way an issue is 'packaged' to encourage certain interpretations and discourage others. Framing can shape how issues are conceptualised and ultimately interpreted. A framing determines what knowledge can be legitimately called on, what metrics are used to capture issues and what action is taken to address them. It is a concept derived from sociology and also used in psychology and economics, where it explicitly links to decision-making. Importantly, framing also relates to the interests of various actors who promote a course of action or who support a particular decision. Narratives, storylines or caricatures might be used to exemplify or justify a particular approach – to 'back-up' the frame. But framings matter because of the way in which they relate to these interests and actors, and the way that they support claims and particular actions, as becomes apparent in the in-depth analysis in following chapters. So whilst acknowledging the diversity of meanings and perspectives associated with the term resilience, the following discussion highlights three framings that are especially important for the application of resilience concepts to international development.

Resilience as the new sustainability

The report of the high level panel on sustainable development, prepared to advise the UN Director General in advance of the Rio+20 meeting on sustainable development in 2012 was titled 'Resilient People, Resilient Planet: A Future Worth Choosing'.[5] It offers the possibility that sustainable development is being redefined in terms of resilience; it puts resilience at the very heart of sustainability. This book explores these possibilities and the ways in which new thinking on resilience is being applied, and how these ideas could potentially transform thinking on sustainable development.

There is an uneasy relationship between resilience and sustainability. Much has been written dissecting sustainable development and sustainability; and the burgeoning literature on resilience indicates a similar move towards a critical literature. But are the concepts linked, and are they complementary? Do they help us to understand development in new ways?

There are important similarities between them and some similar traits in the way the concepts and terms have been applied and recorded in the literature, summarised in Box 1.4. These are reflected, for example, in earlier debates about sustainable development, widely rehearsed more than two decades ago (notably Redclift 1987; see also Redclift 2006). These earlier discussions highlighted some of the problems with the concept of sustainable development. They observe that it is deliberately vague and slippery, making it extremely difficult to operationalise; that it enables 'green-washing' or green camouflaging (Adams 2008) of current strategies and thus in some sense fosters hypocrisy; that it implies the notion and possibility of sustainable growth. It is perhaps even delusional, distracting attention away from meaningful, more profound change and the root causes of global inequities and environmental degradation (Robinson 2004). A substantial literature has been established debating these and other issues, and indeed some similar arguments have been made recently about resilience.

Charles Redman from Arizona State University has written and researched widely on sustainability and resilience, and he questions the assumptions of complementarity between resilience and sustainability (2014). He highlights the contradictions between the two concepts and makes a case for keeping them distinct in both scholarship and practice. Redman makes his argument by outlining the contrast between adaptation and transformation, which he sees as core concepts respectively in resilience and sustainability science.

Box 1.4 Similarities between sustainability and resilience

- Both are in common usage and have multiple meanings and interpretations.
- Both criticised for being 'malleable', but this is also part of their attraction.
- Both criticised for leading to technical solutions, downplaying the social and political.
- Both often used to promote 'business as usual'.
- 'Mainstreaming' tends to water down aspirations.
- Normative applications are problematic, often contested.
- Both have measurement difficulties.

Source: Author's own

Redman's contrasting elements of resilience and sustainability are summarised in Table 1.3. In line with Leach et al. (2010) he argues that sustainability is about defining future pathways for society in which human well-being is enhanced, social equity is advanced, and environmental integrity is protected. Redman claims that this aligns sustainability more closely with transformation. Resilience on the other hand is not about predicting outcomes, but about building capacity, primarily adaptive capacity to deal with unknown futures. Thus, sustainability prioritises outcomes whilst resilience prioritises process. I further discuss some of these issues, and especially the interplay between adaptation and transformation later in the book – particularly in the final three chapters – as it is key to how resilience might inform transformative approaches to change.

Yet resilience gives us further traction on sustainability. But it also proposes some significantly different things. Sustainability gives a clear message about the integration of social, economic and environmental (and sometimes cultural) dimensions of development. Of course, many of the criticisms concern its emphasis on the environmental, and both sustainability and resilience are observed to privilege a conservationist agenda (Lélé 2013). But sustainability also necessitates explicit consideration of both intra- and inter-generational equity, about equity in the present between rich and poor, and between current and future generations, in a far more explicit way than a resilience approach does (Lélé 1998). But resilience deals more overtly with change, and with uncertainty and disequilibria than sustainable development, which might be crudely interpreted to present change in more linear ways, without the emphasis on thresholds or disturbance.

Proponents of sustainability science see resilience as a component of or contribution to sustainability. Sustainability science has emerged in the past decade as an integrative research framework, and is described as a 'field of

Table 1.3 Contrasting elements of resilience and sustainability

Resilience theory approach	Sustainability science approach
Change is normal, multiple stable states	Envision the future, act to make it happen
Experience adaptive cycle gracefully	Utilise transition management approach
Origin in ecology, maintain ecosystem services	Origin in social sciences, society is flawed
Result of change is open ended, emergent	Desired results of change are specified in advance
Concerned with maintaining system dynamics	Focus is on interventions that lead to sustainability
Stakeholder input focussed on desirable dynamics	Stakeholder input focussed on desirable outcomes

Source: Redman 2014: 37

research dealing with the interactions between natural and social systems, and with how those interactions affect the challenge of sustainability: meeting the needs of present and future generations while substantially reducing poverty and conserving the planet's life support systems' (Kates 2011: 19449). Sustainability science thus aims to bridge scientific and disciplinary boundaries, work across scales and integrate different styles of knowledge in engaged scholarship. It sets out to bring concerns for sustainability to the centre of science and policy. It is underpinned by four branches of science – biological, geophysical, social and technological – but in many ways, sustainability science has become a blanket term for a field of interdisciplinary analysis of environment and development issues.

But whether or not the two concepts of resilience and sustainability are linked – complementary, contradictory or distinct scientifically – they are used in similar ways in the rhetorics of public policy around international development. In some ways resilience *has* become the new sustainability – because it is applied across many different arenas (for example, in adaptation to climate change; Brown 2011) and as a normative good, just as its predecessor 'sustainable' was. So resilient becomes a signifying descriptor, as sustainable is or was – for example in 'resilient growth', or 'climate-resilient development'. This is explored in greater depth in the next chapter analysing policy discourses and applications (including measurement) of resilience, and in Chapter 5 looking at ideas about 'sustainable adaptation'.

Resilience might add to our understanding of sustainability by providing some more dynamic views on what a transition to sustainability may mean. In developing concepts and measure of thresholds and feedbacks for example, it provides insights into dynamic change rather than the simplistic trajectories of change assumed in earlier writings on sustainable development. Benson and Craig (2014) rather provocatively posit 'The End of

Sustainability', suggesting resilience thinking as providing a more appropriate footing for meeting the challenges of the 'Anthropocene' including large-scale and irreversible change, approaching thresholds and uncertainty and non-linearity of change. They contend that 'sustainability' has failed, citing the outcomes of Rio+20 as evidence. Yet resilience may be able to rise to these challenges and provide critical insights for development in a dynamic, uncertain world, but only if we rethink and extend some of its current applications.

Boundaries and limits

The notion of planetary boundaries is rapidly becoming a centrepiece of research on resilience and sustainability, and in many ways a defining metaphor for sustainable development. It is also a key component of a significant framing of resilience around the scale of human activities and their impact on a finite planet. It has become highly visible both in scientific and policy arenas.

The planetary boundaries concept was originally presented in a paper published in the journal *Nature* in 2009, co-authored by Johan Rockström from Stockholm Resilience Centre and 28 co-authors (Rockström et al. 2009a). To date it has been cited more than 2000 times (according to Google scholar in June 2014) and the longer version, published in *Ecology and Society* the same year (Rockström et al. 2009b) more than 630 times. The concept of planetary boundaries itself has become influential and prominent in discussions around global governance, sustainability and development – the most frequent claim being its influence on the Rio+20 agenda and the discussions around the Sustainable Development Goals (Stockholm Resilience Centre 2012). However, this is not without controversy, evidenced by the discussions and defence of the concept on the Stockholm Resilience Centre website and the debate at the Resilience 2014 conference about the framing of planetary boundaries and assumptions about global governance and the politics of planetary boundaries, and potential privileging of global environmental concerns over local social ones (see Melissa Leach's reflections on this[6]). The planetary boundaries analysis has recently been updated (Steffen et al. 2015) where the authors suggest it provides a framework for a new paradigm that 'integrates the continued development of human societies and the maintenance of the Earth system in a resilient and accommodating state' (Steffen, 2015, abstract).

The planetary boundaries framework proposes that a set of critical Earth system processes together maintain the planet in relatively stable Holocene-like conditions. It is these conditions which have allowed humankind to thrive and develop – they enabled settled agriculture and the expansion of human activities and populations. The analysis identifies the critical Earth system processes and their dynamic interactions at local, regional and global scales and proposes boundary levels which avoid key tipping points or biophysical thresholds to define a 'safe operating space' for humankind. The iconic diagram produced by this research shows the nine boundaries associated with chemical pollution, climate change, ocean acidification, stratospheric ozone depletion, nitrogen and phosphorus cycles, global freshwater use, change in land use, biodiversity loss and atmospheric aerosol loading. It shows that for perhaps three of these earth system processes – climate change, biodiversity loss, and nitrogen and phosphorus loss – the boundaries are estimated to already have been exceeded. The boundaries are not fixed limits as such. These Earth system processes are now being seriously undermined by these human activities and are in danger of crossing critical thresholds which could – if transgressed – move the planet away from the Holocene-like (relatively) stable and benign conditions. Of course there are many caveats, conditionalities and uncertainties associated with calculating the boundaries and thresholds upon which they depend.

The influence of the planetary boundaries framework and its rapid uptake in policy and think tanks is reflected in initiatives such as the Planetary Boundaries Initiative,[7] which describes itself as 'a legal think-tank advocating effective governance of Earth-system processes', and claim that 'planetary boundaries is a ground-breaking concept', and that 'the science of planetary boundaries provides a powerful new vision for future human development'.

The key ideas and concepts underpinning the planetary boundaries analysis integrate thinking about limits, Earth system science, sustainability science, and are framed within resilience thinking that understands the global processes as part of a complex self-regulating system, which has multiple states and which is characterised by threshold effects and feedbacks (see Box 1.5).

The intellectual lineage of planetary boundaries can be traced back to the Limits to Growth model proposed in early 1970s by Denis Meadows and

---

Box 1.5 Planetary boundaries core concepts

- Scale of human activities and the capacity of the Earth to sustain them – drawing on ideas from ecological economics, biophysical constraints, support for human well-being.
- Understanding of essential Earth system processes and the impacts of humans and trajectories of change – from earth system science and sustainability science.
- Resilience framework – using ideas on complex dynamic systems, self-regulating systems, multiple states and threshold effects.

Source: Author's own

Donella Meadows (Meadows et al. 1972; Buttel et al. 1990), positing ecological limits to economic growth. These ideas can be seen to have evolved into understandings of global environmental change in the 1990s and the Earth system science emerging in early 2000s (Adger et al. 2005). Each present global reasoning and view the biosphere as a global system that is undergoing degradation by global-scale processes – be they systemic or cumulative (Turner et al. 1990). They also predict potential pathological scenarios, and each point to the need for globally co-ordinated, multilateral political responses. Rockström et al. (2009b) claim that their approach extends Limits to Growth by incorporating notions such as safe minimum standards and precautionary principle, key concepts which inform sustainable development. A major advance claimed is the focus on the biophysical *processes* of the Earth system that determine the *self-regulating capacity* of the planet. So rather than focusing on the provisioning capacity or sources (as the Club of Rome and Limits to Growth model did) or sinks (as more recent global environmental change and sustainable development conceptualisations do) planetary boundaries is much more about the processes which keep the planet in the Holocene-like conditions of last 10,000 years. Despite fluctuations in rainfall patterns, vegetation distribution and nitrogen cycling, this has remained relatively stable where key biochemical and atmospheric parameters have been fluctuating within a fairly narrow range. Similar to global environmental change perspectives (Turner et al. 1990) that emerged in the 1990s and informed sustainable development thinking, it recognises historic regional and local changes but emphasises that global systemic and cumulative scales change is a relatively recent phenomenon driving significant and perhaps irreversible changes in global processes. Hence, the emergence of the Anthropocene – a new geological epoch where the activities of humankind have become the major source and cause of change in the biosphere.

But in defining a 'safe operating space for humanity' this work has generated criticism on the basis of presumed conflict between global equity and environmental sustainability (Steffen & Stafford Smith 2013). In particular the lack of recognition that global-level environmental goals may constrain development aiming to reduce poverty, and that the planetary boundaries, in aggregating to a global level, overlooks the need for greater access to resources amongst the world's poor. In addition, Steffen and Stafford Smith observe that developed countries suspect that more equity in resource access in a world with finite resources means they must give up some of their material wealth and current consumption. By way of extending the planetary boundaries concept to take account of the needs for development, Raworth (2012) and Leach et al. (2013) propose 11 social boundaries below which lie resource deprivations that endanger human well-being. These 11 social boundaries are based on social issues raised as priorities in more than half of all government submissions to the UN Rio+20 Conference on Sustainable Development, held in June 2012. Combining the inner limits of social boundaries and the outer limits of planetary boundaries, creates a 'doughnut-shaped space within which all of humanity can thrive by pursuing a range of possible pathways that could deliver inclusive and sustainable development' (Leach et al. 2013: 85). Their analysis shows that the social boundaries – the minimum standards for sustainable development – have not been reached for each of the social issues.

This perspective raises the question of how far the planetary boundaries concept can go in informing sustainable development. Steffen and Stafford Smith (2013) for example, make a strong defence of planetary boundaries as a *starting point* which might lead to a disaggregated analysis taking spatial social equity considerations into account. This recognises regional differences and suggests possible management options at international scale, for example, technology transfer to reduce emissions of aerosols, to reduce pollution and to improve water quality and reduce demand (see Steffen & Stafford Smith 2013: Table 1). But 'negotiating these boundaries' requires 'navigating alternative pathways' in the terms of Leach et al. (2013) in order to find a safe and just operating space for humanity, and is an intensely political issue. Leach et al. observe that there are likely to be many different pathways, across different scales and they will be aligned with different cultures and contexts, different visions, and values and each will have different risks, costs and distributions of power and costs and benefits between different social actors. They will each have different implications and different outcomes in terms of social justice. So their definition and the choices made about them – their adjudication that Leach et al. refer to – is intrinsically complex and political, as well as highly dynamic.

So planetary boundaries analysis, the concept itself, arises out of the application of resilience ideas within a framework of a broader Earth system or sustainability science. But does it, as some have claimed, present the road map for sustainable development? Undoubtedly, the framing of resilience as limits or thresholds within the global Earth system context has very quickly become influential, and the planetary boundaries metaphor itself is very powerful.

Risk and insecurity

We live in uncertain times, and in a highly connected globalised society. A pervasive sense of a world affected by natural and human-made disasters hurtling towards the unknown and the unknowable is propagated in the popular media and public mind. Kuecker and Hall (2011) articulate this as humanity now entering an era 'defined by transition to catastrophic collapse' (19), characterised by multiple, interconnected and large-scale synchronised crises. Hayward (2013), among others, notes that resilience has emerged as an important lens and a policy response in an era of public concern about disasters and risk, including fear of terrorism and environmental catastrophe. Hayward reminds us that 'in our rapidly urbanising world we risk losing sight of a degrading planet pushed to the limits of its capacity to support our growth' (p1).

In this framing of insecurity and new and escalating risks, resilience has growing purchase in the security and international relations domain. Cascio (2009) suggests resilience as 'the next big thing' in *Foreign Policy* magazine's review of emerging trends, highlighting the inability of security to encompass profound uncertainty, non-linearities and surprise in social systems (also Evans et al. 2010). Resilience is increasingly linked to security – and building resilience promoted as a means to enhance security – both to reduce threats and their impacts, and to recover from disturbances.[8] The UK's National Security Strategy (Cabinet Office 2010) specifies resilience as a goal of national security, highlighting resilience of communities both as a goal and as a means to overall security. But the Security Strategy focuses heavily on resilience in the area of civil contingencies – responding to, lessening damage from, and recovering from emergency events such as threats to public health, large-scale fires or explosions, large-scale civil unrest, acts of terrorism, severe weather and flooding – rather than understanding resilience as a property of how systems work at different scales (see also IPPR sponsored *Commission on national security* 2008).

The rise of resilience as a concept in these fields can be seen to be linked to an increased appreciation of multiple stressors and overlapping insecurities which periodically come together to have profound impacts. Furthermore, the global connectedness of environmental and economic systems means that what happens in one place and at one time can very quickly have impacts in other locations. The major interdependencies of climate, food and water insecurity, and their potential convergence into the so-called 'perfect storm' (Sample 2009) was suggested by the then UK Government's Chief Scientific Advisor, John Beddington. It is observed that it is where results of change interact with widespread poverty that the most severe impacts are created (see Box 1.6).

The 'perfect storm' metaphor emphasises the security dimensions of a number of environmental and development issues. It highlights the threats associated with global change processes and impacts, so it is a strong metaphor within a framing of resilience. This linkage of resilience and security is reflected in prevailing research, and the number of research initiatives and centres that have been established in the past decade. This is especially so since the 2007–8 financial crisis; the interlocking concepts of risk, security and resilience have come to the fore in public discourse. This is evidenced in, for example, the number of recent research centres worldwide focusing on security, risk and resilience, demonstrating a prevalent framing of resilience in risk and security applied to different sectors and spheres of contemporary society. This applies the study of resilience and risk to infrastructures, government entities, businesses or the economy, and especially cities. The shift in emphasis to the systems view and towards

---

Box 1.6 A 'perfect storm'

Throughout 2008 and 2009 Sir John Beddington, then UK Government Chief Scientist, used the metaphor of the 'perfect storm' to highlight food, energy and water security in the context of climate change, claiming that the world is heading for major upheavals which are due to come to a head in 2030 (Sample 2009). He pointed to research indicating that by 2030 a whole series of events [will] come together:

> The world's population will rise by 33 per cent from 6bn to 8bn.
> Demand for food will increase by 50 per cent.
> Demand for water will increase by 30 per cent.
> Demand for energy will increase by 50 per cent.

Rapidly increasing urbanisation, prosperity, biofuel production and climate change exacerbate each of these events. The problems combine and reinforce each other to create a 'perfect storm' in which the whole is bigger, and more serious, than the sum of its parts.

This apparently caught the zeitgeist, gained considerable media attention and raised this as a priority in the UK and internationally. It came at a time when intergovernmental negotiations around climate change – in Copenhagen in 2008 – were stalling, and effectively set the tone for ensuing policy-scientific debates such as the Planet Under Pressure conference in the run up to Rio+20.

Source: Author's own

---

integration of a diverse set of risks, and how they might be managed to enhance security and build resilience is apparent.

The US sociologist David Stark reflects on the securitisation and the socialisation of resilience in his essay, titled 'On resilience' (2014). He traces the first systemic studies of resilience in a social system to research after the Second World War on the effects of allied bombing on the German economy, so the securitisation of resilience is not just a recent phenomenon, even if it is a contemporary frame. However, Stark carefully dissects how a risk and security perspective, especially one which relies on probabilistic models of risk, distorts risk perceptions, and in the public consciousness focuses us on large-scale and unlikely disasters. He observes that

> from the first systematic studies of resilience we find an unbroken link of conceptual and methodological developments, moving from the field of security to the field of *securities* and leading to contemporary models of risk – models, moreover, which take into account that other

actors are also using models in attempts to design resilience in response to risky futures.

(Stark 2014: 65)

As he notes, as models become more reliable, the system becomes more unpredictable, yet new technologies promise the 'automation of resilience' – its evolution to predictive science based on probabilistic risk and new information technology including surveillance. Here, the role of the social scientist becomes not only to interject reflexivity about the likely implications and feedbacks of forecasts, but also to be vigilant about the ways in which attempts to mitigate risk and protect against disaster themselves compound risk and introduce new vulnerabilities. Ultimately for Stark 'one of the dangers of security measures is that, in trying to conquer vulnerability, they damage the very processes which are at the basis of resilience' (2014: 68).

This securitising or security framing clearly relates primarily to resilience as resisting shocks or bouncing back, and does not resonate strongly with more dynamic views of resilience either as adaptation or transformation. However, it is extremely influential, especially in areas of planning and urban design, and it speaks very forcefully to financial and business worlds.

Is resilience a buzzword?

Resilience is a prominent term used in everyday discourse to represent the ability of individuals, communities and other systems to respond to trauma. For example, when people talk about their communities after a dramatic event, they inevitably speak of 'resilience'. We see this especially in the way that extreme events are reported and the 'human face' that the media puts on their impact. This is further explored in Chapter 2. Resilience is popularly seen as trait that helps individuals to cope with and recover from trauma.

In January 2013 *Time* magazine declared 'resilience' the buzzword of 2013 (Walsh 2013), indeed its conceptual vagueness and malleability have led others to, often pejoratively, declare it as the latest buzzword. Identifying resilience as a buzzword, implies a temporariness, and a lack of conceptual rigour. Buzzwords get their buzz from being in-words – defining what is in vogue. Yet buzzwords are also distinguished by their ability to have multiple meanings and nuances, depending on who is using them and in what context. In her analysis of development buzzwords, Andrea Cornwall (2007: 472) describes how buzzwords gain their purchase and power through their 'vague and euphemistic qualities, their capacity to embrace a multitude of possible meanings and their normative resonance.' Buzzwords get picked up and applied by different actors to promote their interests to diverse audiences. Their inherent ambiguity and elasticity is one of the key characteristics which enables them to (as

Cornwall notes), 'shelter multiple agendas, providing room for manoeuver and space for contestation' (2007: 474).

This might certainly be the case for resilience as currently articulated and these are all issues I discuss and analyse. This book explores some of these dimensions in how resilience is being used and promoted, and the claims that are made around resilience. The term might certainly be understood as a buzzword in that it clearly resonates with many different actors and interests, and it has powerful normative associations.

In 2013, reporting for the Guardian Global Development Professional Network, Misha Hussein declared 'Resilience is probably the sexiest new buzzword in international development',[9] observing that it gained popularity particularly following the 2008 food, fuel and financial crises. According to this article, the term has assumed such political and financial clout that every funding proposal, every programme, every result must be seen to contribute towards resilience, yet has 'brought utter confusion' – because of its multiple meanings and lack of clear definition and straightforward, operationalisable measurement. Certainly we see a wide range of terms such as resilient livelihoods, resilient development, and climate resilience, in the international development policy literature. 'Resilient' becomes the adjective applied to any action to signal is desirability, its robustness, its sustainability. The following chapter discusses some of these approaches in more detail. Carpenter et al. (2001) discuss how resilience as a metaphor for a healthy and robust system, and as a measurable indicator, are intertwined. They argue that 'although the metaphorical concept of resilience has the power to inspire useful analyses of social ecological systems, much more insight could be gained from empirical analysis, which would require an operational, measurable concept of resilience' (2001: 766). These issues are further explored later; in Chapter 2 analysing recent applications and measures of resilience in international development, and Chapter 4 brings more empirical evidence to understand experiential resilience.

Brand and Jax (2007) identify increased conceptual vagueness with resilience as it moves from its ecological roots. On the one hand this makes resilience a loose 'boundary object' (p8) – malleable, slightly ambiguous and able to draw different interests and actors together; and on the other a descriptive ecological concept. They describe resilience as 'two-faced' (p9). But not only is resilience becoming increasingly vague and normative, its origins as a descriptive concept are being lost, and in Brand and Jax's view, it is increasingly conceived as a perspective or even as a way of thinking applied to social processes such as governance, social learning or leadership, or perhaps as a metaphor for the flexibility of a social ecological system over the long term. Chapter 2 discusses how resilience has been presented as a 'boundary object' whether or not it is able to bring together different scientific fields, or bridge science policy arenas and to produce shared meaning and understanding. The different applications of 'resilience' in international development policy are analysed, and the extent to which these represent new approaches or new thinking, or perhaps re-badging of existing policies with a new buzzword.

Re-framing resilience

A theme running throughout this book concerns the multiple meanings of resilience; the ways these are articulated and interwoven, and in some instances how they are conflated, confused and talk past each other. This depends on who is concerned with resilience and of what and whom, and the need for us to question and understand these applications. Here I develop a pluralist perspective on resilience. The approach melds together different scientific perspectives and different disciplines. A central plank and a starting point is the conceptualisation of resilience in social ecological systems scholarship, which brings an analytical view of linked social and ecological system, or humans and environment as coupled, and notions from complex adaptive systems. But insights from human development fields that give much greater emphasis to human agency are integrated with this to develop a more agency-centred approach which emphasises the socially contingent and dynamic nature of resilience as a property of people and social ecological systems. Importantly this acknowledges and analyses feedbacks and dependencies on ecological and environmental dimensions. Hence it is guided by a political ecology approach.

But it is also crucial to recognise the tensions between a descriptive and normative approach to resilience. This book generally adopts a descriptive approach, understanding that resilience is not always good, but that it is also socially constructed and negotiated. Resilience relates to how people and linked social ecological systems can respond to change – both sudden shocks but other types of changes too. Resilience building is about increasing that capacity, as the Rockefeller Foundation defines it, as

> of an individual, community or institution to survive, adapt, and grow in the face of acute crises and chronic stresses . . . an activity that requires a multi-faceted, interdisciplinary strategy and a systems view to grasp the interconnected and cross-sectoral nature of particularly 'wicked' problems like chronic poverty and global warming.
>
> (Martin-Breen & Anderies 2011: 2)

This seems to be completely appropriate to span development and global environmental change concerns.

In the chapters that follow I build a cogent argument for understanding resilience as a characteristic or property of complex dynamic social ecological systems that can support positive and proactive change. Applying the political ecology approach outlined here, I dissect current understandings, scientific precedents and empirical insights. I build on knowledge and experience of adaptation and add to recent work on transformations across diverse fields. My aim is to thoroughly analyse and revise resilience as a concept that can provide a clear agenda and sound conceptual foundation for development in our era of unprecedented environmental change. Part of this is about re-framing resilience informed by plural views, values and science for development. Partly this is about carefully assessing the political, social and ecological dimensions of change and how they relate to contemporary development priorities.

Having discussed the definitions and meanings, defined the approach and also analysed some current framings, the next chapters analyse the discourses associated with policy and their applications, the different knowledges associated with diverse scientific fields, and the lived experiences of resilience. My re-visioning of resilience expands current understandings and by emphasising a agency-centred approach which builds on the everyday experiences and forms of resilience, and brings three core ideas into the heart of resilience analysis: resistance, rootedness and resourcefulness. I develop these themes in the later chapters.

Notes

1 The Resilience Alliance is a scholarly network made up of member institutions that include universities, government, and non-government agencies. See www.resalliance.org/index.php/key_concepts.
2 Based at Dalhousie University in Halifax, Canada. See http://resilienceresearch.org.
3 *African Health Services*, Special Issue, December 2008 Editorial: Resilience to Disasters: A Paradigm Shift from Vulnerability to Strength. Astier M. Almedom and James K. Tumwine.
4 I refer to this field generally as 'social ecological systems resilience'.
5 www.uncsd2012.org/index.php?page=view&nr=267&type=400&menu=45.
6 http://steps-centre.org/2014/blog/resilience2014-leach/.
7 See planetaryboundariesinitiative.org.
8 Examples include Preparing Scotland: Scottish Guidance on Resilience: www.readyscotland.org/ready-government/preparing-scotland; Resilient US: www.resilientus.org/.
9 www.theguardian.com/global-development-professionals-network/2013/mar/05/resilience-development-buzzwords.

References

Adams B (2008) *Green development: Environment and sustainability in a developing world*. Abingdon: Routledge (3rd edition).

Adger WN, Benjaminsen TA, Brown K and Svarstad H (2001) Advancing a political ecology of global environmental discourses. *Development and Change*, 32(4), 681–715.

Adger WN, Brown K and Hulme M (2005) Redefining global environmental change. *Global Environmental Change*, 15(1), 1–4.

Baggio J, Brown K and Hellebrandt D (2015) Boundary object or bridging concept? A citation network analysis of resilience. *Ecology and Society*, 20(2), 2.

Bahadur AV, Ibrahim M and Tanner T (2010) *The resilience renaissance? Unpacking of resilience for tackling climate change and disasters, SCR Discussion Paper 1. SCR Discussion Paper*, Institute of Development Studies. Available from: http://bit.ly/1cIBONB (accessed 23 October 2013).

Béné C, Newsham A, Davies M, et al. (2014) Resilience, poverty and development. *Journal of International Development*, 26(5), 598–623.

Benson M and Craig R (2014) The end of sustainability. *Society and Natural Resources*, 27(7), 777–82.

Beymer-Farris B, Bassett, T and Bryceson I (2012) Promises and pitfalls of adaptive management in resilience thinking: the lens of political ecology. In: Plieninger, T and Bieling C (eds) *Resilience and the cultural landscape: Understanding and managing change in human-shaped environments*, Cambridge: Cambridge University Press, 283–300.

Blaikie, P (1985). *The political economy of soil erosion in developing countries*. Longman.

Brand F and Jax K (2007) Focusing the meaning (s) of resilience: Resilience as a descriptive concept and a boundary object. *Ecology and Society*, 12(1).

Brown K (2011) Sustainable adaptation: An oxymoron? *Climate and Development*, 3(1), 21–31.

Brown K (2014) Global environmental change I: A social turn for resilience? *Progress in Human Geography*, 38(1), 107–17.

Buttel FH, Hawkins AP and Power AG (1990) From limits to growth to global change. *Global Environmental Change*, 1(1), 57–66.

Cabinet Office (2010) *A strong Britain in an age of uncertainty: The national security strategy*. Available from: http://bit.ly/1eJJDMh (accessed 31 July 2013).

Carpenter S, Walker B, Anderies J and Abel, N (2001) From metaphor to measurement: Resilience of what to what? *Ecosystems*, 4(8), 765–81.

Cascio J (2009) The next big thing: Resilience. *Foreign Policy*. Available from: www.foreignpolicy.com/articles/2009/04/15/the_next_big_thing_resilience (accessed 15 June 2014).

Cornwall A (2007) Buzzwords and fuzzwords: Deconstructing development discourse. *Development in Practice*, 17(4–5), 471–84.

Evans A, Jones B and Steven D (2010) *Confronting the long crisis of globalization: Risk, resilience and international order*. Brookings/CIC report. Available from: http://cic.es.its.nyu.edu/sites/default/files/evans_globalization.pdf (accessed 15 June 2014).

Fabinyi M, Evans L and Foale S (2014) Social ecological systems, social diversity, and power: Insights from anthropology and political ecology. *Ecology and Society*, 19(4), 28.

Fabricius C, Folke C, Cundill G and Schultz L (2007) Powerless spectators, coping actors, and adaptive co-managers: A synthesis of the role of communities in ecosystem management. *Ecology and Society*, 12(1), 29.

Forsyth T (2003) *Critical political ecology: The politics of environmental science*. London and New York: Routledge.

Haddad L (2014) Development horizons: Is 'resilience' a mobilising metaphor for 21st century development? Available from: www.developmenthorizons.com/2014/05/is-resilience-mobilising-metaphor-for.html (accessed 1 July 2014).

Hatt K (2013) Social attractors: A proposal to enhance 'resilience thinking' about the social. *Society and Natural Resources*, 26(1), 30–43.

Hayward B (2013) Rethinking resilience: Reflections on the earthquakes in Christchurch, New Zealand, 2010 and 2011. *Ecology and Society*, 18(4), 37.

Hornborg A (2009) Zero-sum world: Challenges in conceptualising environmental load displacement and ecologically unequal exchange in the world-system. *International Journal of Comparative Sociology*, 50(3–4), 237–62.

IPPR (2008) *'Commission on national security in the 21st century' project*. London: IPPR. Available from: http://bit.ly/1QGThV7.

Kates R (2011) What kind of a science is sustainability science? *Proceedings of the National Academy of Sciences of the United States of America*, 108(49), 19449–50.

Kuecker GD and Hall TD (2011) Resilience and community in the age of world-system collapse. *Nature and Culture*, 6(1; Spring), 18–40.

Lawhon M and Murphy JT (2011) Socio-technical regimes and sustainability transitions: Insights from political ecology. *Progress in Human Geography*, 36(3), 354–78.

Leach M (ed.) (2008) *Re-framing resilience: A symposium report*. STEPS Working Paper 13, Brighton. Available from: http://steps-centre.org/wp-content/uploads/Resilience.pdf (accessed 18 July 2013).

Leach M, Scoones I and Stirling A (2010) *Dynamic sustainabilities: Technology, environment, social justice*. London: Earthscan.

Leach M, Raworth K and Rockström J (2013) Between social and planetary boundaries: Navigating pathways in the safe and just space for humanity. In: Hackmann H and St Clair A (eds), *World Social Science Report 2013 – Changing Global Environments: Transformative Impact of Social Sciences*, UNESCO and International Social Science Council. Available from: http://bit.ly/1GYIE8Z.

Lebel L, Anderies J, Campbell B, Folke C, Hatfield-Dodds S, Hughes TP and Wilson J (2006) Governance and the capacity to manage resilience in regional social-ecological systems. *Ecology and Society*, 11(1), 19.

Lélé S (1998) Resilience, sustainability environmentalism. *Environment and Development Economics*, 3(02), 221–62.

Lélé S (2013) Rethinking sustainable development. *Current History*, 112(757), 311–16.

Lukes S (1973) *Power: A radical view*. Basingstoke and New York: Palgrave Macmillan.

Martin-Breen P and Anderies J (2011) *Resilience: A literature review: The Rockefeller Foundation*. Available from: http://bit.ly/1Hhz2sy (accessed 27 February 2013).

Masten A, Best KM and Garmezy N (1990) Resilience and development: Contributions from the study of children who overcome adversity. *Development and Psychopathology*, 2(04), 425–44.

Meadows DH, Meadows DL, Randers J and Behrens, WW (1972) *The limits to growth*. New York: New American Library.

O'Brien K, Hayward B and Berkes. F (2009) Rethinking social contracts: Building resilience in a changing climate. *Ecology and Society*, 14(2), 12.

Peet R and Watts M (eds) (1996) *Liberation ecologies: Environment, development, social movements*. London: Routledge.

Peterson G (2000) Political ecology and ecological resilience. *Ecological Economics*, 35(3), 323–36.

Peterson G and Hornborg A (2012) A online debate on resilience theory versus political ecology – CSD Uppsala. Available from: www.csduppsala.uu.se/2012/video-and-slides-from-grasping-sustainability (accessed 3 July 2014).

Pritchard L and Sanderson S (2002) The dynamics of political discourse in seeking sustainability. In: Gunderson LH and Holling CS (eds), *Panarchy: Understanding transformations in human and natural systems,* Washington, Covelo, London: Island Press, 448.

Raworth K (2012) *A safe and just space for humanity: Can we live within the doughnut?* Oxford: Oxfam in association with GSE Research, Available from: www.ingentaconnect.com/content/oxpp/oppccr/2012/00000008/00000001/ art00001 (accessed 4 July 2014).

Redclift M (1987) *Sustainable development: Exploring the contradictions.* London and New York: Methuen.

Redclift M (2006) Sustainable development (1987–2005): An oxymoron comes of age. *Horizontes Antropológicos,* Programa de Pós-Graduação em Antropologia Social da Universidade Federal do Rio Grande do Sul., 3(SE), 65–84.

Redman CL (2014) Should sustainability and resilience be combined or remain distinct pursuits? *Ecology and Society,* 19(2).

Robards M, Schoon M, Meek C and Engle, NL (2011) The importance of social drivers in the resilient provision of ecosystem services. *Global Environmental Change,* 21(2), 522–9.

Robinson J (2004) Squaring the circle? Some thoughts on the idea of sustainable development. *Ecological Economics,* 48(4), 369–84.

Rockefeller Foundation (2013) 100 Resilient Cities. Available from: www.100resilientcities.org (accessed 23 April 2015).

Rockström J, Steffen W, Noone K, Persson Asa, Chapin FS, Lambin EF, Lenton TM, Scheffer M, Folke C, Schellnhuber HJ, Nykvist B, deWit CA, Hughes T, van der Leeuw S, Rodhe H, Sörlin S, Snyder PK, Costanza R, Svedin U, Falkenmark M, Karlberg L, Corell RW, Fabry VJ, Hansen J, Walker B, Liverman D, Richardson, K, Crutzen P and Foley JA. (2009a) A safe operating space for humanity. *Nature,* Nature Publishing Group, 461(7263), 472–5.

Rockström J, Steffen W and Noone K (2009b) Planetary boundaries: Exploring the safe operating space for humanity. *Ecology and Society,* 14(2), 32.

Sample I (2009) Beddington: World faces 'perfect storm' of problems by 2030. *theguardian.com.* Available from: www.theguardian.com/science/2009/mar/18/ perfect-storm-john-beddington-energy-food-climate (accessed 12 July 2014).

Stark D (2014) On resilience. *Social Sciences,* 3(1), 60–70.

Steffen W and Stafford Smith M (2013) Planetary boundaries, equity and global sustainability: Why wealthy countries could benefit from more equity. *Current Opinion in Environmental Sustainability,* 5(3–4), 403–8.

Steffen W, Richardson K, Rockström J, Cornell SE, Fetzer I, Bennett EM, Biggs R, Carpenter SR, de Vries W, de Wit CA, Folke C, Gerten D, Heinke J, Mace GM, Persson LM, Ramanathan V, Reyers B and Sorlin S (2015) Planetary boundaries: Guiding human development on a changing planet. *Science,* 347(6223), 1259855.

Stockholm Resilience Centre (2012) Planetary boundaries. Available from: www. stockholmresilience.org/21/research/research-programmes/planetary-boundaries. html (accessed 12 July 2014).

Taylor P and Buttel FH (1992) How do we know we have global environmental problems? Science and globalisation of environmental discourse. *Geoforum,* 23(3), 405–16.

Turner B, Kasperson R and Meyer W (1990) Two types of global environmental change: Definitional and spatial-scale issues in their human dimensions. *Global Environmental Change,* 1(1), 14–22.

Walker BH, Abel N, Anderies J and Ryan, P (2009) Resilience, adaptability, and transformability in the Goulburn-Broken Catchment, Australia. *Ecology and Society*, 14(1), 12.

Walsh B (2013) Adapt or die: Why the environmental buzzword of 2013 will be resilience. *Time*, Washington, DC. Available from: http://ti.me/1ARlrUe (accessed 4 July 2014).

Xu L and Marinova D (2013) Resilience thinking: A bibliometric analysis of socio-ecological research. *Scientometrics*, 96(3), 911–27.

2 Development policy engagement with resilience

Resilience is now widely promoted as a goal of international development policy; this is evident in policy documents and statements in both developed and developing countries, and with respect to many sectors of the economy and society. In these, resilience is presented as an almost uniformly beneficial and desirable trait – of systems, including social ecological systems, the economy, communities, individuals and sub-sectors of society. But resilience is used, applied and interpreted in a number of different ways in these prescriptive arenas. Chapter 1 introduces some of the main framings of resilience, and in this chapter I analyse policy documents on resilience from a range of international agencies, governments, NGOs and think tanks working in international development, environment and related areas. This analysis examines how resilience has been taken up by different policy actors and communities, and the different ways in which it is interpreted, presented and applied.

This chapter identifies three distinct discourses evident across these different policy arenas. One of these discourses in optimistic in outlook, and the other two are pessimistic. But the discourses themselves are often inconsistent, mismatched with scientific thinking, and in some respects, confused. This chapter discusses how, in many cases, resilience ideas are used to support and promote business as usual and not to challenge the status quo, resonating with criticisms of resilience as 'conservative'. But it is also sometimes used strategically to bridge different areas of policy, especially humanitarian and disaster relief with longer-term development investments. It looks at development agencies' increasing search for ways to apply and to measure resilience and discusses whether or not this might provide new insights and innovation for policy and practice in international development.

Policy arenas

In Chapter 1, I reflect on the prominence of the term resilience in the current debates around sustainable development, exemplified in the way in which it was used in discussions around climate change and sustainability. The different framings, for example around security, sustainability and ecological limits, show how resilience has been communicated and popularised. The term has become increasingly prominent in public and policy discourses and debates, especially in the past decade. This chapter probes this phenomenon more deeply and analyses the policy discourses surrounding resilience. It looks at how resilience ideas are articulated and promoted and how they are used to support and advocate particular courses of action, specifically in the field of global change – environmental change and international development. It observes how the measurement of resilience is becoming a focus of work in international development, especially in response to climate change.

Resilience is especially conspicuous in both public and policy discussions that intersect and have particular relevance to global change, including national security, disasters, and environmental change more broadly. Building resilience is promoted as a means to enhance global, national and local security, both to reduce threats and their impacts, and to recover from disturbances. The UK's National Security Strategy[1] (Cabinet Office 2010), for example, specifies resilience as a goal of national security, related to the ability to assess and monitor risks, and to maintain critical infrastructure. Similarly, the term resilience is repeatedly and frequently used in relation to disasters and how individuals, communities and nations are able to respond to shocks and contingencies (see Box 1.1 in Chapter 1). Sometimes, this capacity is related to broader systems factors. For example in responses to the 2010 floods in Pakistan, Patrick McCully, writing in the *Huffington Post*, argues that the catastrophe demonstrates that fundamental systematic change is required in river-basin management worldwide. He states that 'increasing resilience to floods in Pakistan, the US, and just about everywhere else is going to require reversing our river management mistakes through restoring rivers and floodplains, including by taking out embankments and dams' (McCully 2010). The World Food Programme, along with other commentators and NGOs, highlight the resilience of individuals, families and communities to multiple stressors, for example linking flood with conflict (World Food Programme 2010). A strong thread in this field is that 'ordinary' people exhibit resilience in the face of these events in spite of government and other policies which have provoked events or made people and places more vulnerable. Resilience in this context is the ability to withstand and cope, and recover from an event with very little reference to longer-term adaptation or transformational change.

The adoption of resilience ideas and slogans in relation to environmental change resonates strongly with these areas too. Resilience ideas are evident in international science and policy statements such as the *Millennium Ecosystem Assessment* (2005); in the *Human Development* reports from UNDP (UNDP 2008, 2010); World Bank reports (World Bank, 2009a and 2013); the Swedish government sponsored Commission on Climate Change and Development (2009); and initiatives such as the World Bank's pilot for climate resilience (World Bank 2008). Non-governmental organisations including Christian Aid and Oxfam, and think tanks such as the World

Resources Institute have used the term to frame their policy documents and analysis (Brown 2012). Terms such as 'climate change resilience' and 'climate resilient development' have become prevalent, and these are discussed in more detail here.

Various authors and commentators have suggested that the adoption of resilience ideas in international development represents some kind of paradigm shift; however, as Levine and Mosel (2014) point out, it is far too early to judge whether it really has had such long-term or meaningful impact. They identify four main areas where resilience is being talked about and planned. These are, first, new frameworks for resilience building; second, reform of aid bureaucracy to support resilience; third, programming specifically for resilience; fourth, establishing metrics for resilience measurement. This chapter looks especially at how development agencies and other organisations concerned with global change are using resilience to headline their policy and to shape the development of their approach and their programmes. First it examines the shared understandings or meanings – in other words, the discourses – of resilience articulated through policy statements produced by these different agencies. It presents and reviews some of the frameworks that stem from these and then at some examples of their implementation. It then looks at some of the current attempts to measure resilience and evaluate the impacts of resilience-focused policies.

Policy discourses on resilience

A discourse, in its broadest sense, is a common understanding of a phenomenon shared by a particular group of people. Importantly the group of actors adhering to a discourse participate in various ways in producing, reproducing and even transforming it through their oral and written statements (Adger et al. 2001). A discourse may become institutionalised and may be important in shaping activity or directing policy. When resilience first catapulted onto international development and global change policy agenda, I became interested in how different organisations were taking on and articulating the terminology, making claims about resilience, interpreting resilience in different ways partly depending on their context and interests, and how they were beginning to apply these ideas within their organisations and practices. I analysed policy documents across different policy arenas in global change that were, and are, using resilience ideas. I started with a core set of policy documents (shown in Box 2.1) and then have subsequently added to and updated these. I refer to a much broader set of policy documents and grey literature to support this analysis. I first presented these findings in 2010 and they are also included in my chapter in a book edited by Mark Pelling and colleagues, *Climate change and the crisis of capitalism* (Brown 2012). The analysis deliberately includes statements from development, disasters and environmental change, and covers the period 2008 to 2012 – where, as noted in previous chapter, we witness a rapid rise in the use of resilience terms.

Box 2.1 Selected policy documents on resilience

1 UNDP, *Human Development Report 2007/2008*
2 World Bank, *World Development Report 2010*
3 UN Commission on Climate Change and Development (2009)
4 World Bank, Pilot Program on Climate Resilience
5 WRI, *Roots of Resilience* (2008)
6 DFID, *White Paper* (2009)
7 Community and Regional Resilience Initiative, *ResilientUS*
8 US Indian Ocean Tsumani Warning System Program (2007)
9 Christian Aid, Building Disaster Resilient Communities Project
10 DFID, *Defining Disaster Resilience: A DFID Approach* (2011)
11 Christian Aid, *Thriving, Resilient Livelihoods: Christian Aid's Approach* (2012)
12 Montpellier Panel *Growth with Resilience: Opportunities for African Agriculture* (2012)

Source: Author's own

In analysing these documents I take a similar approach to John Dryzek's research on environmental discourses (Dryzek 1997). This examines the text of documents and statements, and identifies the key components of different discourses, based on the language used, and the common assumptions in five different dimensions shown in Box 2.2.

Box 2.2 Analysing policy discourses: five dimensions

1 **Who is resilient and to what?** This identifies basic entities whose existence is recognised or constructed which together constitute the ontology of the discourse and refers to what or whom resilience is applied to, for example ecosystems, humans or social ecological systems.
2 **Assumptions about natural relationships** that define what affects resilience and how, and that may explain how humans and ecosystems are linked and patterns of causation.
3 **Key metaphors** and other rhetorical devices, including narratives, used to explain and promote the discourses.
4 **Agents and their motives,** which refers to who or what are the main actors shaping resilience and who has agency to affect resilience.
5 **Policy prescriptions and normative assumptions** that link specific suggestions of courses of action to promote or manage resilience in the face of change and which are articulated by the discourses.

It is important to recognise that each policy document or statement from an organisation, does not itself represent a single discrete discourse; the discourses might be shared by a number of different documents, or there will be multiple – and sometimes perhaps conflicting – discourses promoted by a single organisation or even within a single policy document. This is demonstrated, for example, in the World Bank's (2008) policy on climate resilience which is discussed later in the chapter. As my analysis shows, the policy documents are sometimes inconsistent and even contradictory in how they define and present resilience. Furthermore, these interpretations often do not correspond closely to scientific understandings of resilience. I explore these scientific meanings in different fields, and then plural narratives about resilience in the next two chapters. So whilst I do not describe each policy or document in detail, I describe the discourses which can be identified by analysing the five components in Box 2.2.

Three discourses are identified by this analysis. The first presents an optimistic view of resilience and global change. It identifies strategies to support, or 'nurture' resilience, using markets and developing market-based mechanisms such as payments for ecosystem services. It views resilience as being entirely compatible with growth-based development strategies – indeed, resilience is itself necessary to defend or protect growth in the face of shocks, perturbations and other changes.

Two other discourses are overtly pessimistic in that they view resilience as necessary to withstand and resist detrimental change and essentially to minimise harm. They differ, however, in terms of their ontologies or stances. One pessimist discourse sees resilience as the ability to withstand externally derived change – so it is closely aligned with disaster risk reduction approaches, where resilience is more narrowly defined in terms of risk reduction.

The second pessimist discourse presents resilience as the opposite of vulnerability, so it is allied to the first pessimist discourse, but sees the social and structural determinants of vulnerability within the system itself. It links resilience with poverty alleviation and views the drivers of poverty as closely coupled to those of vulnerability.

Discourses are important because they reflect a shared understanding and help to shape the policies and approaches taken within agencies. The discourses overlap and reinforce the framings presented in Chapter 1, and they also align with but do not directly or simplistically map onto, different policy arenas – international development, disasters and environmental change. These discourses begin to weave a more nuanced and complex understanding of institutional understandings and uses of resilience ideas, highlighting the way in which resilience is moulded around existing aims and objectives, how it responds to new challenges and incorporates new and already established knowledge, strategies and operations. The elements of the discourses – the five dimensions identified in Box 2.2 – are each now discussed in turn.

Basic entities: resilience of what?

Characterising the basic entities in a discourse involves identifying how resilience is applied and how the system is defined in the text of policy documents. In these documents, resilience is applied to a number of different entities. *Ecological, social* and *economic resilience* are identified, and resilience is seen as a property of communities (Christian Aid 2007; WRI 2008), the rural poor (WRI 2008; Christian Aid 2012), countries and states (World Bank 2009a), as well as of infrastructure and even enterprise. Some documents recognise resilience as a system property and refer to communities as systems for example, or to linked social ecological systems.

Development agencies tend to focus on resilience as a characteristic or property of livelihoods (see DFID, WRI, Oxfam and Christian Aid). Resilient livelihoods enable households to withstand shocks and maintain well-being and in the longer term to escape poverty. DFID's resilient livelihood approach builds directly on its sustainable livelihoods framework (see Ashley and Carney 1999) and puts an emphasis on assets as providing the base for resilient livelihoods and the means of sustaining a household. Figure 2.1 reproduces Christian Aid's diagram showing the components that make up their definition of a thriving, resilient livelihood. This shows a resilient livelihood to be made up of a range of different components, which includes resources or assets people can draw on; their approach to risk and adaptiveness; broader aspects of health and well-being; and their power and voice.

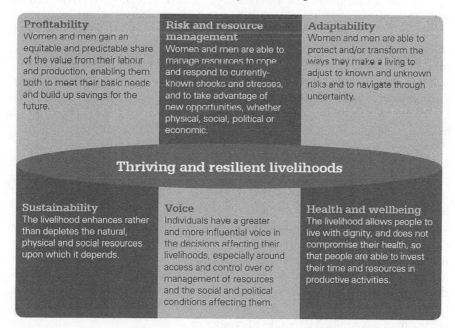

Figure 2.1 Christian Aid's components of thriving and resilient livelihoods

Source: Christian Aid 2012: 2

Rather than a focus on livelihoods or communities, for the Montpellier Panel (2012) – a group of agricultural and development experts convened prior to the Rio+20 meeting – resilience is 'the capacity of agricultural development to withstand or recover from shocks and to bounce back to previous level of growth' (p11). Resilience therefore has to be strengthened in order for growth to continue, a view that is strongly echoed in World Bank's view of climate resilient development discussed in subsequent sections here.

Assumptions about relationships

What affects resilience and how are these relationships described in the policy documents? The WRI 2008 report *Roots of resilience* highlights resilience as a property of the rural poor that determines how they are able to respond to environmental and social challenges including climate change, loss of traditional livelihoods, political marginalisation, breakdown of customary village institutions. Christian Aid (2007) on the other hand, sees resilience as a property associated with responding to disasters, within a context where disasters are increasing in frequency and impact – they refer to increased vulnerability to natural hazards. Each of the policy documents view climate change as a significant threat to resilience, whether acting in synergy with, or in addition to, other important drivers of change.

The UK Department for International Development (DFID) identifies four elements that make up its approach to resilience – its 'resilience framework' is reproduced in Figure 2.2. This shows how resilience is viewed as being context and disturbance specific. It presents resilience as primarily made up of three components – exposure, sensitivity and adaptive capacity. In this sense, as these three variables are conventionally seen to determine vulnerability, resilience is indeed 'the other side of vulnerability' (DFID 2011: 9). In the DFID view these capacities are determined primarily by assets, and will influence how a system (household, community, etc.) responds to a particular disturbance. In the policy statement itself resilience building has a central (though not exclusive) focus on assets. The assets and resources can be social, human, technological, physical, economic, financial, environmental, natural and political, and are correlated to the sustainable livelihoods approach. The framework offers four reactions to disturbance, which include bounce back and bounce back better, thus it spans the capacity to survive, cope, recover, learn and transform, reflecting a more multi-dimensional understanding of resilience. So it recognises resilience as a characteristic associated with a range of different responses, and to different types of disturbance, much in line with social ecological systems resilience. It also signifies the necessity to specify resilience of what to what, rather than assuming a generalised resilience.

Importantly most of these policy documents recognise multiple stressors, and a number note the potential synergistic impacts of these – either

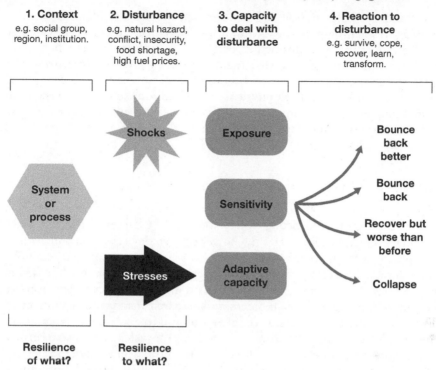

1. Context	2. Disturbance	3. Capacity to deal with disturbance	4. Reaction to disturbance
e.g. social group, region, institution.	e.g. natural hazard, conflict, insecurity, food shortage, high fuel prices.		e.g. survive, cope, recover, learn, transform.

Shocks

Exposure

System or process

Sensitivity

Stresses

Adaptive capacity

Bounce back better

Bounce back

Recover but worse than before

Collapse

Resilience of what?

Resilience to what?

Figure 2.2 DFID's four elements of a resilience framework

Source: DFID 2011: 7

simultaneously or sequentially. Thus Christian Aid (2012: 3) sees interconnected challenges requiring integrated solutions. Almost uniquely amongst this set of documents, Christian Aid puts an emphasis on underlying drivers, including power imbalances, as well as climate change, in the context of the daily reality of poor peoples' lives and the hazards that threaten them. It highlights the importance of assets, but presents them differently to DFID, with an emphasis on assets providing buffers for poor people. Of all the statements, Christian Aid's 2012 document is the most overtly political in presenting inequality and lack of power and voice as core elements undermining resilience.

Likewise the Montpellier Panel (2012) recognises the interactions between multiple stressors, identifying seven threats that constitute major challenges for food security. These are high food prices; the concentration of malnourished people in sub-Saharan Africa; the impacts of malnourishment particularly for women and children; increasing populations which necessitate doubling of food production; environmental degradation; fuel and fertiliser price increases; and climate change. It promotes resilient agricultural growth as a means to overcome this, defining this as agricultural growth which produces enough food, is accessible to all, and is able to withstand multiple stresses and shocks. The report articulates a security framing, resonating with the idea of a 'perfect storm' discussed in Chapter 1. The report makes a distinction between stresses and shocks but again highlights interlinkages between them, and cites growing insecurity and threats: 'Because the planet is becoming more densely populated and increasingly urbanised, both physical and social interactions are becoming more complex and fast moving. As a consequence minor adverse events become amplified and threats to agricultural growth are multiplying in frequency and scale' (p11).

Metaphors and narratives

Metaphors and other rhetorical devises such as narratives are used to frame, illustrate and justify particular policy stances. In Chapter 1, the 'planetary boundary' metaphor illustrates the power of such devices to encompass and encapsulate a set of complex ideas. In these policy documents, for example, WRI develops a set of narratives to support its emphasis on 'ecosystem services based enterprises' which focuses on markets for ecosystem services as a means to support smallholders and research managers. A number of different metaphors and narratives are employed to explain the dynamics of change and to justify the idea of nurturing ecosystem-based enterprises, using expressions such as 'Green Livelihoods' and 'Turning Back the Desert'. The document presents a set of narratives which provide optimistic interpretations of resource conservation and natural resource management such as from the Sahel, illustrating that with appropriate support and enabling institutions and incentives, smallholders can achieve greater security, productivity and stem environmental degradation.

Christian Aid's document (Desai & Moss 2007) presents a quite different pessimistic narrative of socially differentiated vulnerability. It explains how Hurricane Mitch in Central America undermined household resilience, producing pronounced gender-differentiated impacts. It illustrates this by highlighting the increased incidence of domestic violence after the hurricane: 'men found it difficult to cope with the intense pressure of trying to provide for their families and coping with loss with little or no psycho-social support' (p3). Similar to other disasters literature it makes a direct link between a community-wide shock, and individual and personal well-being and mental health issues. In turn this has further impacts on families and households. Interestingly, the document itself makes little mention of resilience apart from in its title.

In DFID's document (2011) the desired response to a shock or stress is to 'bounce back better' (p9) where the system capacities are enhanced or sensitivities and exposures are reduced, leaving a system that is better able to deal with future shocks or stresses. The report uses the example of the Zambezi Floodplain Management programme in Mozambique to illustrate how persistent flooding has forced communities to diversify livelihoods and change cropping patterns (with technical support and funding), which has in turn strengthened resilience to drought and other impacts of climate change. Similar narratives – smallholder benefits associated with agricultural development, technical assistance and livelihood diversification – are used in many development and climate change policy documents, and certainly resonate with the accounts presented by WRI, World Bank, the Montpellier Panel and Oxfam.

Agents: who affects resilience?

A wide range of agents are identified as the key players affecting resilience and who have responsibility and agency. They reflect in part of course the main interests and stakeholders of the institutions producing and promoting the policies.

Governments are often identified as key agents in the policy documents. For example, the World Bank emphasises the role of government in integrating climate risk into national development planning. It also sees the opportunity for learning between agents – mainly confined to cross-national sharing of lessons at country, regional and global scales. The Montpellier Panel clearly states its recommendations are aimed at governments and European donors working in partnership with local and international private sector actors, NGOs and Civil Society Organisations. Communities are emphasised by Christian Aid, reflecting the focus on community resilience noted in broader academic literature on resilience (Brown 2014).

Normative assertions

The prescriptions offered in these documents are quite diverse, but they do have important features in common. These include the concepts of scaling up, capacity development and disaster risk reduction. WRI's view of building resilience involves 'scaling up' – enhancing local-scale projects and funding – and also necessitates drawing people into markets, especially markets for ecosystems services, as a means to deliver what they refer to as the 'resilience dividend' to the rural poor. They define this as a 'community driven model to manage their ecosystem assets and build them in enterprises can experience a marked increase in their resilience' (WRI 2008: 27). The core idea is that markets and particularly payments for ecosystem services provide positive case studies of the 'power of self-interest and community ownership, the enabling value of intermediary organisations and how communication and networks can provide new ideas and support' (WRI 2008: 3). Yet Watts (2011) cites the WRI document as articulating a form of green governmentality, where resilience represents 'a brave new world of turbulent capitalism and the global economic order, and a new ecology of rule' (p88) and is used as a means of justifying a set of policies and actions which further commodification and capitalist development.

The enterprise model presented includes both adaptation and mitigation of climate change (for example by drawing smallholders into carbon markets), and overall presents the potential for positive outcomes and opportunities from finding innovative responses to change. In this respect it relates strongly to prescriptions from other agencies (e.g. Food and Agriculture Organization [FAO], in its *State of Food and Agriculture* 2007) that promote payments for ecosystem services as a means of alleviating rural poverty and simultaneously addressing environmental concerns. Scaling up and mainstreaming are also core features of the prescriptions suggested by DFID.

In contrast to the WRI's rather optimistic view of resilience as enabling and building capacity, the Commission on Climate Change and Development (2009) uses resilience in a quite different way and presents a different set of prescriptions. The Commission views resilience as much more concerned with managing and reducing risks – it resonates here with security framing discussed in Chapter 1. It has a specific focus on the poorest, and presents resilience in terms of capacities, closely aligned with disaster risk reduction, and the need for greater knowledge and communication. It stresses the need to build resilience of communities, countries and regions to cope with unexpected events. They emphasise that the aim is 'not to strive to foresee the unforeseeable but to train ourselves to cope with it … not to clarify, map and plan for every single surprise, but to train to be surprised' (Commission on Climate Change and Development, 2009: 77).

Disaster risk reduction has taken on new meanings and has been given renewed momentum for many development agencies in the face of climate change. Christian Aid (2007) emphatically states that climate change will increase the number and impact of various types of disasters around the world. The emphasis of its policy, exemplified in the Building Disaster Resilient Communities Project, is on disaster risk reduction; integrating early warning systems and linking local, sub-national and national systems. They prioritise the need to 'climate proof' development and humanitarian work. Again though, it is not clear what is novel or innovative in the resilience approach in terms of activities and actions. For example, they stress the need for strengthening water and watershed management in response to 'new' dangers from climate change, manifest in slow onset disasters such as drought, yet these strategies have long been advocated.

Some of these themes resonate with the rhetoric and policies of the World Bank, through its 2010 World Development Report (World Bank 2009b), and specifically its objective of climate resilient growth. This aims to address climate change risks within project design, and to integrate risk and build capacity in developing countries. It again links adaptation and mitigation, and sees integration into markets as a key means of achieving this. Thus resilience is used – as in the WRI document discussed above – to promote existing agendas concerning expanding markets, strategies which may generate new vulnerabilities and inequalities. Raising agricultural productivity, building knowledge infrastructure and providing energy for all are also core elements. The Pilot Program for Climate Resilience (PPCR) under the Strategic Climate Fund (African Development Bank Group 2008) is a major means of implementing these objectives, with specific regional strategies developed (e.g. the Strategy for sub-Saharan Africa, World Bank 2009a). As discussed in detail in the following section, the emphasis here is on making current development activities resilient to climate change, but the main focus is still on growth, productivity and markets. In a later World Development Report, *Risk and opportunity: Managing risk for development* (2013) the Bank claims that effective risk management can improve resilience to negative shocks and the ability to take advantage of positive shocks, recognising that shocks might provide 'windows of opportunity' and a chance to innovate and change for the better.

The Montpellier Panel, as DFID, World Bank and others, stresses not only the technical innovations, scaling up and intensification of good practice, but also the role of markets at different scales. Building enabling environments, reducing price volatility and facilitating flows of private investment (see also World Bank, WRI and DFID) are emphasised to support 'resilient markets'. As shown in Figure 2.3, their prescription involves three dimensions: markets, agriculture and people, which are interdependent. Importantly each demand political will – again emphasising the role of governments and donors as key agents.

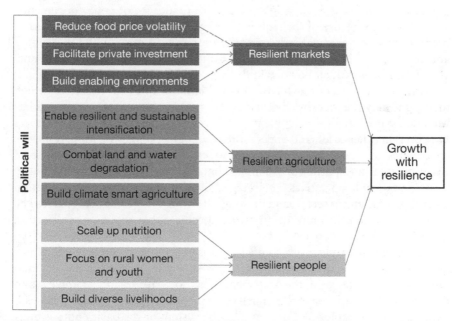

Figure 2.3 The Montpellier Panel's agricultural growth with resilience

Source: The Montpellier Panel 2012: 6

Resilience is seen as an important integrating concept in many of these policy statements – as a means of integrating climate change adaptation and mitigation (World Bank) or to bring together disaster response and poverty alleviation agendas (Christian Aid). DFID refers to resilience as an 'integrating concept' (p5) and a means to build bridges across its priority policy concerns of climate proofing, resilient growth and conflict and fragility. It refers to cross-fertilisation happening through integrating disaster risk reduction, climate change adaptation and social protection strategies of resilience building, for example in Rwanda, inferring a synergistic or win-win outcome is possible. Likewise, Christian Aid talks of resilience as a bridging concept and as a means of finding integrated solutions to interconnected problems. However Christian Aid's rhetoric is overall less optimistic, noting that uncertainty means that a 'solution that works today may need to be completely overturned tomorrow' (p3). Overall very little of the policy literature refers to the possibility of trade-offs inherent in resilience approaches (Robards et al. 2011).

Table 2.1 summarises the examples of policy prescriptions from the policy documents that were analysed. Of course, they are applying resilience in quite different contexts; so they are not compared, but they represent elements of different discourses, and illustrate the diversity of interpretations of resilience in the development and environment field. As Chapter 1 shows, resilience has become a very prominent and very powerful term in these policy communities in recent years. But in reviewing these and a broader set of policy documents, an evolution in the use of the term is evident. Initially the term was applied to promote business as usual, and to justify current and growth and market-led approaches. This is most evident in the examples from WRI, Montpellier Panel and World Bank, which I discuss in further detail in the next section. Rigg and Oven (2015) document this for rural Asia, defining a 'liberal resilience' which they argue plays into a growth-development-resilience 'trap', where growth has become a synonym for development, and development a synonym for resilience. However, in other cases, and under more recent critical interrogation, resilience contributes to more genuine attempts to change policy to take account of environmental and particularly climate change, and to merge different policy communities and imperatives. This is especially true of the disaster risk reduction and humanitarian aid sector, and the longer-term development perspective. The next sections look in more detail at the specific applications of resilience in development and aid policy.

There are multiple and conflicting discourses in each of the policy documents, and importantly there is inconsistency in how terms are used and ideas applied. Many of the statements do not capture the complex and dynamic systems approach that much of resilience scholarship encompasses (I explore this in Chapter 3). For example, the system or the basic entity is most often poorly defined – only the WRI document (2008) links the environmental and social in any 'systems' conceptualisation or framework (this mentions a 'Social Ecological System'). Most of the other documents refer to a poorly

Table 2.1 Summarising policy prescriptions on resilience

How resilience is defined	*Prescription/focus*
US Indian Ocean Tsunami Warning System Program (2007) *How resilient is your coastal community?* **Community resilience** is the capacity to adapt to and influence the course of environmental, social and economic change.	Coastal resilient communities take deliberate action to reduce risk from coastal hazards with the goal of avoiding disaster and accelerating recovery in the event of a disaster. They adapt to changes through experience and applying lessons learned (p3–1).
Christian Aid (Desai & Moss 2007) *Overexposed: Building disaster-resilient communities in a changing climate* Resilience is the **ability to withstand the impact of shocks and crises**. It is determined by people's assets and their ability to access services provided by external infrastructure and institutions.	The Building Disaster Resilient Communities Project supports local partner organisations … to strengthen communities' abilities to manage and recover from crises and to prepare for and reduce the risk of future disasters (p8).
Oxfam (2009) *People-centred resilience* Resilience is the **ability of a joint social and ecological system – such as a farm – to withstand shocks, coupled with the capacity to learn from them and evolve in response to changing conditions.**	Makes the case for investing in building up the resilience of vulnerable farming communities as a critical stepping stone to addressing the global challenges of food security, climate change adaptation and climate change mitigation (p7). Building resilience involves creating strength, flexibility and adaptability.
Commission on National Security in the 21st Century (2008) *Shared responsibility* Resilience is not specifically defined – it is the **opposite of vulnerability and is a character of critical infrastructure,** enterprise and communities.	A commitment to building national resilience, especially in our infrastructure, by measures including educating and increasing the self-reliance of communities is an integral part of security policy (p4).
World Bank (2008) *Pilot program for climate resilience* Resilience not specifically defined, but aim to '[integrate] **climate risk and resilience into core development planning,** while complementing other on-going activities'.	Aim to provide incentives for scaled-up action and transformational change in integrating consideration of climate resilience in national development planning consistent with poverty reduction and sustainable development goals (section B paragraph 4).
The Montpellier Panel (2012) *Growth with resilience: Opportunities in African agriculture* Resilience is the **capacity of agriculture development** to withstand or recover from stresses and shocks, and then bounce back to the previous level of growth.	The challenge is to generate agricultural growth that produces enough food, ensures it is accessible to all, is inclusive of the most vulnerable and is resilient, by creating resilient markets, resilient agriculture and resilient people (p5).

(Continued)

Table 2.1 continued

How resilience is defined	Prescription/focus
DFID (2011) *Defining disaster resilience: A DFID approach paper* **Disaster resilience** is the ability of countries, communities and households to manage change by maintaining or transforming living standards in the face of shocks or stresses … without compromising their long-term prospects (p6).	Resilience is seen as an important integrating core concept which can build bridges across disaster risk reduction, climate change adaptation and poverty alleviation, recognising multiple stressors including financial crisis and climate change; views resilience as 'the other side of vulnerability' (p9) and focuses on mainstreaming resilience and on building assets.
Christian Aid (2012) *Thriving, resilient livelihoods: Christian Aid's approach* A **resilient livelihood** is one that enables **people** to anticipate, organise for and adapt to change – good or bad, slow or sudden (p2).	The resilient livelihoods framework recognises interconnected challenges – multiple stressors – and the need for integrated solutions which recognise underlying drivers of poverty and vulnerability, power imbalances, climate change and the daily reality of poor peoples' lives.

Source: Author's own

defined entity. Other core concepts, including thresholds, feedbacks, networks, connections, transformation and transformative change are almost completely absent.

The areas where there is stronger correlation to the conceptual or theoretical literature are in some of the approaches to promoting resilience through applying adaptive management and in disaster risk reduction. But the problems identified by Bahadur et al. (2010), namely the contestations around meanings and relationships with vulnerability and adaptive capacity for instance, and the lack of clear implications for operationalisation, are evident. One manifestation is the tendency to promote resilience in order to maintain some form of stability – as WRI articulates: 'increased resilience results in ecosystem stability, social cohesion and adaptability, economic enterprise' (WRI 2008: 6), and the need to scale up the resilience of the poor to accommodate environmental and social change, which is a very passive notion of responding to change. The emphasis on integration into the economic mainstream, into cash economies and markets is quite striking. The core objective of the World Bank and the Montpellier Panel on making growth and current development resilient in the face of climate change shows how resilience is seen as a means to continue business as usual. The central programmatic focus on climate resilient development is now explored in more detail to understand how resilience is being applied in development strategies and investment, and to further dissect these contestations and tensions in conceptualisations and operations.

Interrogating climate resilient development

The importance of resilience as an integrating or bridging concept is a theme I return to at various points in this book. But as Levine and Mosel (2014) ask, what has resilience meant in practice? Here I look in much more detail at the way in which one agency, the World Bank, uses the concept of resilience to bridge development and climate change agendas, and the extent to which risk reduction, climate change adaptation and mitigation are brought together in meaningful ways through their strategies for climate resilient development. I further develop the discussion in the next chapter, which analyses whether resilience is really a boundary object as some analysts have claimed.

The World Bank has been one of the most prominent among the proponents of climate resilient development as a policy discourse and as a set of implementable actions. As part of this strategy, the World Bank has developed a programme to pilot and demonstrate ways to integrate climate risk and resilience into core development planning while supplementing and bolstering its on-going activities. The Pilot Program for Climate Resilience (PPCR)[2] is part of the Strategic Climate Fund (SCF), a multi-donor Trust Fund within the Climate Investment Funds, overseen by the World Bank. The overall objective of the programme is to provide incentives for scaled-up action and transformational change in integrating consideration of climate resilience in national development planning consistent with poverty reduction and sustainable development goals. So the rhetoric is about positive change and synergy between poverty and sustainable development.

Approved in 2008, the projected 'pledged resource envelope' has increased from $614 million (September 2009) to $975 million (July 2010), and as of 31 March 2012, $800 million had been endorsed through 13 strategic programmes ($460 million as grants and $340 million as near-zero interest credits) (Climate Funds Update 2015). This has been pledged by eight country donors: Australia, Canada, Denmark, Japan, Germany, Norway, UK and USA. There are nine countries and two regions which have been invited to participate, including: Zambia, Mozambique, Bolivia, Bangladesh and regions in the Pacific and the Caribbean. Participation is contingent upon recipient countries fulfilling the criteria of the respective trust funds, that is, adopting Bank and donor conditions in exchange for financing. For the PPCR, eligible countries will have to submit 'country investment strategies' which will be assessed by the Strategic Climate Fund PPCR Sub-Committee. Priority is given to highly vulnerable Least Developed Countries eligible for Multilateral Development Bank concessional funds, including the Small Island Developing States.

In the PPCR it is development itself – the process of wealth generation – which is being made more resilient to the impacts of climate change. The PPCR is designed to provide finance for climate resilient national development plans and to build on existing National Adaptation Programs of Action (NAPAs). It provides only short-term funding and the purpose is to distil lessons over the next few years that might be taken up by countries, the development community, and the future climate change regime, including the Adaptation Fund.

Its objectives are summarised as:

- pilot and demonstrate approaches for integration of climate risk and resilience into development policies and planning;
- strengthen capacities at the national levels to integrate climate resilience into development planning;
- scale up and leverage climate resilient investment, building on other on-going initiatives;
- enable learning-by-doing and sharing of lessons at country, regional and global levels.

Resilience themes, including capacities and learning are evident here, but again the emphasis is on 'mainstreaming' climate risks into current development operations. This is echoed in the World Bank's strategy for climate resilient development in Africa (World Bank 2009a) that spells out what it means by climate resilient development in more detail. It centres on making adaptation a core component of development, with a particular focus on sustainable water resources, land, and forest, integrated coastal zone management, increased agricultural productivity, health problems and conflict and migration. It aims to focus on knowledge and capacity development by improving weather forecasting, water resources monitoring, land use information and disaster preparedness; investing in appropriate technology development; and strengthening capacity for planning and co-ordination, participation and consultation. It emphasises the benefits from mitigation opportunities through access to carbon finance against land use changes and avoided deforestation, promoting clean energy sources (e.g. hydropower) and energy efficiency, and adopting cost effective clean coal energy generation and reduced gas flaring. It also highlights the opportunities of 'scaling up financing', in other words maximizing flows of these conditional investments. This strategy reinforces dominant development agendas based on economic growth and technology transfer. But these strategies themselves may have inherent biases – for example they may benefit some sections of society over others – and may further reinforce existing vulnerabilities, or bring new ones and do not address root causes of vulnerability. These aspects come to the fore in discussion of climate change adaptation and development in Chapter 5.

These suggested climate change and development policies are *not* radical. They reflect core values and development paradigms that call on market environmentalism, ecological modernisation and environmental populism. Overall the proposals encourage an incremental rather than a radical approach. Here, resilience is part of a strategy that mainstreams climate change adaptation and mitigation into development efforts. Climate change is understood as a challenge to current development but also as providing opportunities. These opportunities emphasise the potential of new investments, for example through carbon markets and market-based mitigation. This is especially the case for Africa in terms of land use, forest management and renewable energy. Yet these approaches have been widely criticised as new means of appropriating resources, ('Green Grabbing' according to Fairhead et al. 2012) and undermining local rights or creating new vulnerabilities (Beymer-Farris & Bassett 2012). But the World Bank's strategy suggests that adaptation is 'fundamentally about sound, resilient development' (World Bank 2009a: 5), and resilience in this context is bestowed by mainstreaming climate change, by making adaptation and risk management core development elements, by fostering knowledge and capacity (particularly in terms of access to information and forecasting) and scaling up financing. There are explicit links made to disaster preparedness and disaster risk reduction, to providing layers of insurance protection and safety nets where appropriate, and to building 'climate smart' systems.

In this way the approach mixes a number of different aspects of resilience thinking, including multiple and cross-scale dynamics, an emphasis on shocks and disturbances to the system; but also aspects of engineering-resilience, i.e. withstanding and resisting shocks. Importantly, climate resilient development is not presented as anything fundamentally different to current development – it emphasises that current plans need to be 'climate-proofed'; in other words, the potential future impacts of climate change and associated risks must be built into present plans. But the premise of continued growth and the benefits of this strategy are not questioned; according to this approach, climate change reinforces the need to do development 'better', more effectively and with emphasis on shifting vulnerabilities and how they may re-configure the distribution of costs and benefits within society. This mirrors discussions on climate change adaptation, where distinctions can be made between approaches which see adaptation as necessary to protect development, and those which see climate change and adaptation as an opportunity to change development (for example Ensor & Berger 2009), discussed more fully in Chapter 5.

But how have PPCR projects actually been implemented? Ayers et al. (2011) present an interesting account of the PPCR process in Nepal that provides insights into the application of these ideas and how they are intertwined with existing adaptation and development interventions and investments. New terms, such as 'climate resilience' and 'transformational change', are interpreted differently by different stakeholders, including the Nepal Ministry of Environment, consultants and donors. The key aspect of this analysis is that climate resilience was seen as synonymous with adaptation by some stakeholders, but distinct by others. Ayers et al. conclude that the remit of the PPCR has not been consistently defined by the different stakeholders in Nepal and has been used to justify a number of different and not necessarily consistent approaches. The interpretations and actions reflect both the vested interests as well as underlying assumptions of the different stakeholders – but negotiating and re-negotiating the objectives and projects under PPCR entails negotiating ownership, and reflects current or pre-existing resource and power allocations.

An important finding from Ayers et al.'s study is that the need to demonstrate short-term results militates against more structural and fundamental institutional and governance change, which might be seen as necessary for transformation. Within the programme itself this plays out as a reluctance to hand over power to the national government. Overall the incentives are to 'climate proof' existing investments rather than look for innovations and new opportunities, ultimately mirroring conventional adaptation funding emphasising 'technocratic solutions and externally driven expert judgement in defining and managing risk' (p78). This is at odds with a more dynamic and transformative view of resilience. Ayers et al. suggest that the combination of underlying political economy and power dynamics, and programme funding that needs to demonstrate results, means that these 'progressive concepts' are negotiated and remoulded at a national level to fit more conventional framings and dove-tail to existing projects.

But a whole plethora of different terms are now applied by development agencies – 'climate resilience development', 'climate smart development' to address climate change. ODI (Mitchell & Maxwell 2010) maps these and introduces the term 'climate compatible development' positioned at the intersection of conventional development strategies, climate change adaptation and climate change mitigation; and 'climate resilient development' as an integration of development and adaptation strategies. Climate resilience development thus defined is therefore perhaps more about adapting than transforming conventional development practice.

Levine and Mosel (2014) observe that resilience frameworks promoted by aid organisations have sometimes been misleading, and have confused the relationships between humanitarian aid and development. Furthermore, they imply unreasonable possibilities as goals (e.g. the idea that resilient people become better off the more shocks they suffer), and they fail to address the underlying roots of vulnerability and inequality. Levine and Mosel comment that the frameworks or conceptualisations create a description for resilience that is too far removed from reality to provide any kind of feasible guide for working in the real world. Reviewing these (selective) approaches to operationalising resilience, it is difficult to see that they represent a new paradigm for development or even any significant change in how development policy gets implemented. They adopt a new lexicon and some new concepts, but still do not address fundamental causes – of poverty, inequality, global change or vulnerability. Later chapters further discuss this and suggest how resilience concepts might underpin a more fundamental shift in development policy. The next section examines some of the recent attempts to operationalise and measure resilience in development. If resilience becomes a core objective of policy, then how is progress towards it measured, and the effectiveness of investments by international donors, governments and NGOs assessed?

Pinning it down: measuring resilience

Despite frequent observations that resilience is unmeasurable, many organisations and scholars are currently grappling with 'pinning down' resilience. An interesting document produced by Mercy Corps (2013) observes that the evidence for informing resilience programming remains 'woefully thin'. Despite multiple frameworks and the rhetorics of development agencies promoting resilience, they argue that 'if investments to strengthen resilience are to be effective they must be informed by more rigour and critical analysis, for whom and to what' (p1). Many commentators and agencies have suggested that in order for resilience ideas to be operationalised then metrics are necessary, and since 2012 particularly, there has been a number of efforts to develop measures of resilience that can be applied to monitor and evaluate development interventions. The logic is that once investments are made in building resilience, then measures against objectives are necessary.

Thus we observe DFID developing a set of complex indicators for monitoring and evaluation of projects funded under its Building Resilience and Adaptation to Climate Extremes and Disasters (BRACED) programme (DFID 2014), so that projects can report against a Key Performance Indicator, specified under a LogFrame, the project management tool used by DFID. Indicators are developed to calculate the number of people with improved resilience within a beneficiary population. This then constitutes an outcome indicator, measuring the resilience of individuals and how this has changed as a result of an intervention, assessing the relative change in resilience, rather than an absolute level of resilience. This raises many technical and conceptual problems: attributing change to a project, adapting an indicator for context and project specifics, distinguishing between specific and general resilience (resilience to a specific disturbance, or system resilience, see Folke et al. 2010), and the challenge of quantitatively measuring a set of perhaps latent capacities. The guidance to measure the number of people whose resilience has been improved as a result of project support suggests nine necessary steps outlined in Box 2.3, producing a complex – and presumably expensive – procedure.

Similarly, other major aid agencies and NGOs are developing methods for measuring resilience, ranging from the highly technical and quantified, to more participatory and qualitative, many mixing methods. For example, Reaching Resilience,[3] a consortium of NGOs and researchers aiming to better integrate knowledge and practice on disaster risk reduction, climate change adaptation and poverty reduction has produced an online Resilience Assessment handbook for aid practitioners and policy makers (Reaching

---

Box 2.3 Measuring resilience the DFID way

1   Identify beneficiaries, shocks and stresses and their consequences;
2   Develop a project theory of change;
3   Identify factors affecting resilience that the project is expected to influence;
4   Develop indicators of resilience;
5   Establish how to identify unexpected consequences;
6   Develop a sampling methodology;
7   Calculate numbers of individuals with improved resilience measured by indicators relevant to project activities and outputs;
8   Attribution – estimate numbers with improved resilience as a result of the project;
9   Report numbers with improved resilience as a result of the project.

Source: Adapted from DFID 2014

Resilience 2013). But each agency has different emphases, different objectives and different approaches. The DFID methodology is important because many of the NGO projects are funded under the BRACED programme through the UK International Climate Fund, a major investment programme for climate change and development. I do not review each of the different approaches and measures here, but rather highlight key initiatives and the way they integrate and apply resilience concepts, and what the measures might mean in the context of the uptake of resilience by development agencies.

Notable amongst these moves, FAO has spearheaded work on resilience related to food security and has initiated a number of attempts to measure resilience, including a proposal for a 'Resilience Tool' (Food and Agriculture Organization [FAO] concept note, n.d.) building on pioneering conceptual and empirical work by Alinovi and colleagues for the EC-FAO Food Security Programme (Alinovi et al. 2009) which shows the origins and explanation of the Resilience Index. This work is strongly rooted in resilience science and concepts; it understands the food system as a complex system made up of interacting social and ecological components, working across multiple dimensions at different scales. Specifically they observe that as a stochastic evolutionary system, i.e. one that behaves randomly, a food system displays path dependency, discontinuous changes, multiple equilibria and non-linearity. This itself represents an important departure from more simplistic and static perspectives on food security which emphasise food production. The conceptual model builds directly on resilience assessment proposed by Walker et al. (2002), presenting four steps defining: Resilience of what? Resilience to what? Resilience assessment; Resilience management. It sees resilience measurement as a component of vulnerability analysis, where vulnerability to food insecurity depends on a household risk of exposure, and its resilience to such risks. The risks are in most cases unpredictable and thus it is difficult to measure vulnerability itself.

Resilience here is defined as 'the ability of a household to keep within a certain level of well-being (i.e. being food secure) by withstanding shocks and stresses' (FAO concept note nd: 1). It will depend on the livelihood options available to the household and on how effectively it can manage risks, but this definition implicitly encompasses ex-ante actions that reduce household exposure to risk of food insecurity, and ex-post actions which enable a household to cope after a crisis. Six factors are combined into an index that gives an overall quantitative resilience score. These factors are: income and access to food; assets; social safety nets; access to basic services; household adaptive capacity; and the stability of these factors over time. Table 2.2 suggests some of the available indicators for these factors.

Table 2.2 Common indicators for the FAO resilience model

Component	Indicators
Income and food access	• Average per person daily income (local currency/person/day) • Average per person daily expenditure (local currency/person/day) • Household food insecurity access score • Dietary diversity and food frequency score • Dietary energy consumption (kcal/person/day)
Access to basic services	• Physical access to health services (ordinal, 1 to 3) • Quality score of health services • Quality of educational system (ordinal, 1 to 6) • Perception of security (ordinal, 1 to 4) • Mobility and transport constraints (ordinal, 1 to 3) • Water, electricity and phone networks (count)
Social safety nets	• Amount of cash and in-kind assistance (local currency/person/day) • Quality evaluation of assistance (ordinal, 1 to 4) • Job assistance (binary yes/no response) • Frequency of assistance (number of times assistance was received in the last six months) • Overall opinion of targeting (assistance targeted to the needy; to some who are not needy; or without distinction)
Assets	• Housing (number of rooms owned) • Durable index (Principal Component Analysis on list of items: TV, car, etc.) • Tropical Livestock Unit (TLU) equivalent to 250 KG • Land owned (in hectares)

(Continued)

Table 2.2 continued

Component	Indicators
Adaptive capacity	• Diversity of income sources (count, 0 to 6) • Educational level (household average) • Employment ratio (ratio, number of employed divided by household size) • Available coping strategies (count, 0 to 18) • Food consumption ratio (share of food expenditure divided by total expenditure)
Stability	• Number of household members that have lost their job (count) • Income change (ordinal; increased, the same, decreased) • Expenditure change (ordinal; increased, the same, decreased) • Capacity to maintain stability in the future (ordinal, 1 to 5) • Safety net dependency (share of transfers on the total income) • Education system stability (ordinal; quality increased, the same, decreased)

Source: Food and Agriculture Organization of the United Nations (n.d.): www.fao.org/docrep/013/al920e/al920e00.pdf (accessed 4 July 2014). Reproduced with permission

Whilst the resilience scores and the radar charts used in the FAO publication to visualise them as the relationships between different components or for different locations or households of different characteristics, are robust (they have been rigorously tested and validated), they do require extensive empirical data. They are designed to use as much existing data as feasible. For example, they are based on the Living Standards Measurement Surveys (LSMS) which many countries regularly collect, but they will ultimately depend on the quality and reliability of those data. The FAO's concept note proposes that once a baseline is established then rapid information gathering methods, based on participatory approaches, can be used to update and supplement, for example when a particular crisis ensues, so that the resilience score can be used to directly inform and target interventions. The 'resilience score' then is not just a tool for monitoring and evaluation, but one to identify, target and shape policy and investment.

Most of these metrics measure resilience at a household or individual scale, using a set of externally generated project specific indicators. They provide before and after measures and they acknowledge the difficulties in capturing the dynamics and processes behind a 'latent property'. The study by Béné et al. (2011) takes a more participatory bottom-up approach to apply resilience analysis to small-scale fisheries in West Africa and focuses on identifying thresholds in a more dynamic approach. Their paper conceptualises small-scale fisheries as complex systems, characterised by a degree of irreducible uncertainty, likelihood of non-linear change and emergence. They use the language and concepts from social ecological systems resilience, but construct a participatory diagnostic approach to identify and analyse possible states (the extent to which well-being needs are met), trajectories of change (better, worse, no change, uncertain) and critical thresholds in the system. Thresholds are analysed across four domains – people and livelihoods; governance and institutions; the natural system and external drivers – to explicitly account for multiple dimensions and scales in the social ecological system. This analysis highlights important potential trade-offs. Particularly in the context of acute poverty, there may be a trade-off between provision of basic needs and fundamental rights such as food security, and the system resilience. This can be seen as a manifestation of a trade-off between specific and system resilience, as discussed by Folke et al. (2010). This emphasises that too narrow a focus on one system level and one set of drivers can easily erode the resilience of the system at different scales or to other drivers (Adger et al. 2011). This issue is discussed in more detail in Chapter 5, but the key question concerns what a priority for poverty alleviation or elimination means for this trade-off between specific and general resilience. A number of methods have also been proposed for measuring community or social resilience (see for example Buikstra et al. 2010; Maclean et al. 2014; Norris et al. 2008) and indeed NGOs have applied community scale assessments (e.g. Transition Network and Carnegie Trust and others 2011). Similarly the links between a multi-dimensional well-being approach and resilience are discussed by authors such as Coulthard (2012), and Armitage et al. (2012).

So whilst development agencies in particular seek quantitative measures of resilience in order to be able to assess progress against policy and intervention objectives, many NGOs and analysts recommend more participatory and qualitative approaches to assessing resilience. Levine (2014) summarises five approaches to measuring resilience, shown in Table 2.3. He maintains that the myriad interpretations – and wide appeal – of resilience as a concept makes measurement problematic, and that many agencies get overly-focused on the technical details of quantification, rather than considering *why* they need to measure resilience. No single tool is appropriate to measure resilience, in part because of its multiple applications.

A number of conceptual and practical challenges arise in trying to pinpoint and measure resilience. These relate to identifying resilience *of whom* or *what*, and the tendency to conflate people and place. Secondly, specifying resilience *to what*, where it is critical to avoid confusing general and specific resilience. Third, *when* to measure, as assessing past resilience to an event, or even present resilience, may not reflect resilience in the future. If resilience is about a set of characteristics or capacities, some of which may be latent and only come into play in a given set of circumstances, then an objective, *a priori* assessment is problematic. Finally, the lack of relevant and accurate data is also a hindrance to quantifying resilience.

Table 2.3 Five approaches to measuring resilience

Quantification based on functionality	Applied in computer systems, infrastructure and engineering where performance can be assessed related to specific functions of a system.
Quantification based on indicators and characteristics	Development of indices currently favoured by many development and aid agencies, e.g. FAO. Most use judgement to derive indicators of generic characteristics of resilience. Assume that improvements in any of the indicators improve overall resilience.
Quantification based on food access	Based on existing models of Household Economy Approach, focusing on the household economy and limited to the economic opportunities or options. It does not attempt to include or predict wider institutional or political determinants of risk or vulnerability. Treats resilience as an outcome measure linked to food security, food access or well-being.
Quantification based on activities	Measures changes in peoples resilience based on different investments or interventions, so compares at aggregate level the differences in outcomes of different interventions (e.g. building capacity prior to an event versus support during or after). This approach has very limited applications.
Quantification derived from theoretical resilience frameworks	Few of the suggested measures of resilience are derived from theoretical frameworks on resilience, but that at present the frameworks do not really help in finding appropriate measures, and do not identify which factors are likely to help different people cope with different difficulties in different circumstances.

Source: Adapted from Levine 2014

These factors lead Levine (2014) to comment that rather than investing greater effort and more resources in developing a definitive measurement, organisations and scholars should step back and ask *why* they need a measure. He suggests that, rather than searching for a universal measure, efforts should be focused on 'ensuring we get better at quantifying the things that really matter' (p17). Additionally, there needs to be strengthening of existing good practice around analysis, assessment and monitoring, and making sure organisations can respond to these exercises. Impact monitoring should inform management of interventions, and steps taken so that organisations learn over time and that comparative assessments inform investment choices, which are accountable. Importantly, understanding the determinants of resilience in different contexts is more important than trying to quantify the outputs of resilience-focused interventions.

These increasing moves to measure, and the demands for common interpretations and definitions of resilience may be linked to its status as a boundary object. Star and Griesemer (1989) point out that boundary objects acquire status through distinct interpretive communities, enabling them to transcend core differences in interpretation for the purposes of alignment required to perform particular work, and developing methodological standardisation and measures are a key aspect of this. Green (2010), for example, analyses 'participation' as a boundary object in international development, remarking that while participation acquires boundary object status as a fairly abstract term, adding a technical dimension (through standardisation and methods) enhances its tangibility. The boundary object – participation or resilience – thus becomes formalised within a set of practices, procedures and measures. This also leads to routinisation, and a closing off of more innovative approaches to problem solving. Analysis in Baggio et al. (2015) reveals that we might regard resilience as a boundary object in that it is interpreted by different groups in different ways, and attempts at measurement might also indicate its status as boundary object. But there is no agreed formula, procedure or technique. And perhaps as Levine (2014) points out, these efforts are incompatible with an approach that gives emphasis to the process of building resilience rather than its specific pre-determined outcomes.

New development paradigm or business as usual?

Folkema et al. (2013: 1) write that 'resilience is now at the heart of development thinking, climate change adaptation and humanitarian policy', and they explain how resilience thinking might change the way development agencies do development. Chapters 6 and 7 further develop this theme and present a new vision of resilience for development. But, as this chapter shows, as yet there is little evidence of *profound change* – merely early indications of a shift perhaps.

The UNDP Human Development Report for 2014 is entitled *Sustaining human progress: Reducing vulnerabilities and building resilience* so it clearly puts resilience at the centre of the development agenda. Likewise we have seen how major multilateral agencies – the World Bank, EU – and bilateral aid agencies and NGOs are adopting ideas and terminology around resilience. We can observe that resilience is – as Brand and Jax (2007) and others have suggested – a boundary object in so far as it can bring together different interests and stakeholders through a set of loosely linked ideas. It has also been used strategically as a bridging concept – for example DFID talks of it as a concept to link disaster response and longer-term development. Each of the agencies uses resilience slightly differently – giving different emphasis and providing different normative suggestions for policy and action. For many, resilience is a slogan – or a buzzword – the new term used to garner support; for others it has become a technical challenge, and a novel way to operationalise and measure the impact of development assistance. These applications and the sophistication and level of detail differ across not only the different agencies but also the different development arenas. Where resilience is applied in a generalised way (as in UNDP for example) and as a core element of development to reduce vulnerability to multiple known and unknown threats, building resilience becomes the objective of development. In other instances resilience is seen more strategically as necessary to address specific threats to food security and to respond effectively to climate change and new macro economic risks. I look in more detail at the potential contributions of resilience analysis to mainstream development in Chapter 6 and to climate change adaptation in Chapter 5.

The *Human Development Report 2014* exemplifies one approach to integrating resilience into mainstream development. It talks about resilient human development and defines human resilience as 'ensuring that peoples' choices are robust, now and in the future, enabling people to cope and adjust to adverse events' (p5). It sees resilience as a set of general capacities, so that 'resilience is about ensuring the state, community and global institutions work to empower and protect people' (p5). But it also talks specifically about resilient growth. It presents four guiding principles to reducing vulnerability and enhancing resilience (Box 1, p5): embracing universalism; putting people first; committing to collective actions; co-ordinating between states and social institutions. It can thus be seen that UNDP applies resilience very much to its own core interests and thematic priorities; human rights, multilateral co-ordination and of course human development.

Commenting on the efforts of US government and major corporate partners such as Microsoft, IBM and Monsanto to address food and water security, big data and sustainability, Emily Gertz (2014), writing in the *Guardian*, questions whether these approaches to climate change resilience represent anything new. Levine and Mosel (2014) observe that many resilience programmes actually constitute a re-phrasing or re-labelling of efforts.

The current applications of resilience concepts in development policy, often in response to climate change, but also prompted by economic crises and perceived threats to food, water and energy security – closely aligned with a security framing – represent, to date, minor adjustments to conventional development strategies. Many of the policy documents reviewed emphasise defence of the status quo. In each of them and in the discourses they represent, resilience is applied in a normative sense, and overwhelmingly to resist or accommodate change, and to enhance stability. Rarely is resilience seen as part of more transformative or dynamic responses. Many of the prescriptions and the programmes which flow from them are technocratic and managerial, which Brooks et al. (2009) suggest is inevitable given the dominant modernist development paradigm. Furthermore, other commentators and scholars see resilience as the latest in a line of concepts used to further the ideology of neoliberalism and markets: just as Watts (2011) reflects on WRI's strategy, McMichael (2009: 252) sees 'climate proofing' as a 'new profit frontier' and that the responses of the multilateral development agencies to climate change seek to 'marketise' development and adaptation. Similarly Rigg and Oven's (2015) growth-resilience-development trap, is articulated in how policy orthodoxies around economic growth and market integration are bolstered by a 'liberal resilience'.

Jerneck and Olsson (2008) analyse resilience as one of the dominant discourses surrounding adaptation to climate change. They identify the implicit normative assumption of the preservation of the system and thus relate resilience to resistance to change. Hence, the need for the ecological system to remain within its 'basin of attraction' (Walker et al. 2004) has implications for the social system. They argue that there is a built-in contradiction in the concept of resilience when it is applied to complex systems where sub-systems with conflicting goals are linked. They maintain that resilience might be useful for guiding adaptation, for instance, in systems where there is no inherent conflict between social and ecological components of systems. But they contend that a resilience approach, exemplified by insurance-based approaches as promoted by the World Bank, might actually delay a long-term solution to vulnerability by encouraging people to continue with a livelihood that is not sustainable in the longer term. But the approach they promote fails to take into account multiple stressors and system-wide change over different spatial, social and temporal scales, so does not incorporate central tenets of resilience science (see Chapter 1). This is further explored in Chapter 5. But Jerneck and Olsson take a narrower view of resilience, and argue that ideas of resilience underline recovery more than fundamental change (2008). Yet this view of resilience – which is shared by others – means that resilience applied in development and climate change fields is more likely to lead to incremental rather than profound change – as this is the case for policy discourses analysed here.

These conceptual contradictions, and what they mean for a resilience approach are discussed by Cannon and Muller-Mahn (2010). They contend that a resilience approach translates into a scientific and technical approach akin to 'imposed rationality' that is alien to the practice of ordinary people. They argue that resilience is de-politicised and does not take account of the institutions within which practices and management are embedded, and this is also evident in Ayers et al.'s (2011) analysis of implementation of PPCR. They argue that these problems stem from the origins of resilience in systems thinking, and in transferring a concept derived from the analysis of natural systems to social phenomena. But in a political economy appraisal, Nasdasdy (2007) provides an explanation of why, despite a rhetoric stressing management for resilience that might be at the expense of short-term stability, there are powerful interests working against such a dynamic or adaptive strategy. He argues that capitalist production demands a permanent degree of short-term stability and so long as this system remains, the pressure to make management decisions based on the stability of one or two key resources will remain. But for McMichael too the fundamental problem is with the neoliberal development paradigm, and so what is required is an ontological break with the standard market episteme of the development project and its particular 'global ecology' (2009: 259). All this implies that resilience is not currently challenging this paradigm, but is used and applied to further it. But this could also underscore and inform more radical departures and strategies to address multiple and profound change. The following chapter explores some of these issues by reviewing the theoretical and conceptual roots of resilience in different fields of science.

Notes

1 The Commission on National Security in the 21st Century, 2008 was a two-year commission, co-chaired by Lord Paddy Ashdown and Lord George Robertson, to conduct a detailed assessment of the changing global security environment and the specific challenges and opportunities this poses for Britain: www.ippr. org/security.
2 The Pilot Program for Climate Resilience (PPCR): www.afdb.org/en/topics-and-sectors/initiatives-partnerships/climate-investment-funds-cif/strategic-climate-fund/pilot-program-for-climate-resilience-ppcr.
3 www.reachingresilience.org.

References

Adger WN, Benjaminsen TA, Brown K, and Svarstad H (2001) Advancing a political ecology of environmental discourses. *Development and Change*, 32(4), 681–715.

Adger WN, Brown K, Nelson DR, Berkes F, Eakin H, Folke C and Tompkins EL (2011) Resilience implications of policy responses to climate change. *Wiley Interdisciplinary Reviews: Climate Change*, 2(5), 757–66.

African Development Bank Group (2008) Pilot Program for Climate Resilience (PPCR). Available from: http://bit.ly/1F5DBWh (accessed 21 October 2013).

Alinovi L, Mane E and Romano D (2009) *Measuring household resilience to food insecurity: Application to Palestinian households.* ESA Working Paper. FAO, Agricultural and Development Economics Division, EC-FAO Food Security Programme. Available from: http://bit.ly/1Fdqi7A (accessed 11 December 2014).

Armitage D, Béné C and Charles A (2012) The interplay of well-being and resilience in applying a social-ecological perspective. *Ecology and Society*, 17(4), 15.

Ashley C and Carney D (1999) Sustainable livelihoods: Lessons from early experience. Available from: www.eldis.org/vfile/upload/1/document/0902/DOC7388.pdf (accessed 12 May 2015).

Ayers J, Kaur N and Anderson S (2011) Negotiating climate resilience in Nepal. *IDS Bulletin*, 42(3), 70–9.

Baggio J, Brown K and Hellebrandt D (2015) Boundary object or bridging concept? A citation network analysis of resilience. *Ecology and Society*, 20(2), 2.

Bahadur AV, Ibrahim M and Tanner T (2010) *The resilience renaissance? Unpacking of resilience for tackling climate change and disasters, SCR Discussion Paper 1.* Institute of Development Studies, Brighton. Available from: http://bit.ly/1cIBONB (accessed 23 October 2013).

Béné C, Evans L, Mills D, Ovie S, Raji A, Tafida A, Kodio A, Sinaba F, Morand, P, Lemoalle J and Andrew N (2011) Testing resilience thinking in a poverty context: Experience from the Niger River basin. *Global Environmental Change*, 21(4), 1173–84.

Beymer-Farris B and Bassett T (2012) The REDD menace: Resurgent protectionism in Tanzania's mangrove forests. *Global Environmental Change*, 22(2), 332–41.

Brand F and Jax K (2007) Focusing the meaning(s) of resilience: Resilience as a descriptive concept and a boundary object. *Ecology and Society*, 12(1), 23.

Brooks N, Grist N and Brown K (2009) Development futures in the context of climate change: Challenging the present and learning from the past. *Development Policy Review*, 27(6), 741–65.

Brown K (2012) Policy discourses of resilience. In: Pelling M, Manuel-Navarrete D, and Redclift M (eds), *Climate change and the crisis of capitalism: A chance to reclaim, self, society and nature*, London: Routledge, 37–50 .

Brown K (2014) Global environmental change I: A social turn for resilience? *Progress in Human Geography*, 38(1), 107–17.

Buikstra E, Ross H and King C (2010) The components of resilience—Perceptions of an Australian rural community. *Journal of Community Psychology*, 38(8), 975–91.

Cabinet Office (2010) *A strong Britain in an age of uncertainty: The National Security Strategy.* Available from: http://bit.ly/1eJJDMh (accessed 31 July 2013).

Cannon T and Müller-Mahn D (2010) Vulnerability, resilience and development discourses in context of climate change. *Natural Hazards*, 55(3), 621–35.

Christian Aid (2007) *Christian Aid's approach to building disaster resilient communities*, Christian Aid, London.

Christian Aid (2012) *Thriving, resilient livelihoods: Christian Aid's approach. Christian Aid Briefing*, London. Available from: http://bit.ly/1PkZFEl.

Climate Funds Update (2015) Climate Funds Update. Available from: www.climatefundsupdate.org/listing/pilot-program-for-climate-resilience (accessed 21 October 2013).

Commission on Climate Change and Development (2009) *Closing the gaps: Disaster risk reduction and adaptation to climate change in developing countries*, Stockholm, Sweden.

Commission on National Security in the 21st century (2008) *Shared Responsibility*, final report of the IPPR. Available from www.ippr.org/publications/shared-responsibilitiesa-national-security-strategy-for-the-uk.

Coulthard S (2012) Can we be both resilient and well, and what choices do people have? Incorporating agency into the resilience debate from a fisheries perspective. *Ecology and Society*, 17(1), 4.

Desai B and Moss S (2007) *Overexposed: Building disaster-resilient communities in a changing climate*. Christian Aid. Available from: http://bit.ly/1IwCFy9.

DFID (2009) *Eliminating world poverty: Building our common future*. White Paper, DFID London. Available from www.gov.uk/government/uploads/system/uploads/attachment_data/file/229029/7656.pdf.

DFID (2011) *Defining disaster resilience: A DFID approach paper*. London: Department for International Development.

DFID (2014) *Methodology for reporting against KPI4*, Christian Aid, London. Building Resilience and Adaptation to Climate Extremes and Disasters (BRACED) programme. Available from: http://bit.ly/1K4ZbhT.

Dryzek JS (1997) *The politics of the earth: Environmental discourses*. New York: Oxford University Press.

Ensor J and Berger R (2009) *Understanding climate change adaptation: Lessons from community-based approaches*, Practical Action, Rugby. Available from: http://practicalaction.org/publishing/understanding_climate_change_adaptation (accessed 4 July 2015).

Fairhead J, Leach M and Scoones I (2012) Green grabbing: A new appropriation of nature? *Journal of Peasant Studies*, 39(2), 237–61.

Folke C, Carpenter SR, Walker B, Scheffer M, Chapin T and Rockström J (2010) Resilience thinking: Integrating resilience, adaptability and transformability. *Ecology and Society*, 15(4), 20.

Folkema J, Ibrahim M and Wilkinson E (2013) *World Vision's resilience programming: Adding value to development*, World Vision, London. Available from: http://bit.ly/1E53v8I (accessed 13 May 2015).

Food and Agriculture Organization (FAO) (2007) *The state of food and agriculture 2007*, FAO, Rome. Available from: www.fao.org/docrep/010/a1200e/a1200e00.htm (accessed 11 March 2015).

Food and Agriculture Organization (FAO) (n.d.) *Measuring resilience: A concept note on the resilience tool*, FAO, Rome. Available from: www.fao.org/docrep/013/al920e/al920e00.pdf.

Gertz E (2014) White House partners with Amazon, Microsoft and others on climate change resilience. *Guardian*. Available from: www.theguardian.com/sustainable-business/2014/aug/01/obama-food-resilience-climate-change-data-microsoft-amazon-monsanto (accessed 12 May 2015).

Green M (2010) Making development agents: Participation as boundary object in international development. *The Journal of Development Studies*, 46(7), 1240–63.

The Intergovernmental Coordination Group for the Indian Ocean Tsunami Warning and Mitigation System (2005) The IOC Tsunami Programme. Available from: http://bit.ly/1L0zAED (accessed 15 June 2014).

IPPR (2008) *Commission on national security in the 21st century project*. IPPR, London. Available from: http://bit.ly/1QGThV7.

Jerneck A and Olsson L (2008) Adaptation and the poor: Development, resilience and transition. *Climate Policy*, 8(2), 170–82.

Levine S (2014) *Assessing resilience: Why quantification misses the point*. London: ODI. Available from: http://bit.ly/1mo2tSe.

Levine S and Mosel I (2014) *Supporting resilience in difficult places: A critical look at applying the 'resilience' concept in countries where crises are the norm*. Humanitarian Policy Group Commissioned Report, London: ODI. Available from: http://bit.ly/1QGYKLt.

Maclean K, Cuthill M and Ross H (2014) Six attributes of social resilience. *Journal of Environmental Planning and Management*, 57(1), 144–56.

McCully P (2010) Global lessons from the Pakistan flood catastrophe. *Huffington Post*. Available from: www.huffingtonpost.com/patrick-mccully/global-lessons-from-the-p_b_691928.html (accessed 14 July 2014).

McMichael P (2009) Contemporary contradictions of the global development project: Geopolitics, global ecology and the 'development climate'. *Third World Quarterly*, 30(1), 247–62.

Mercy Corps (2013) *What really matters for resilience? Exploratory evidence on the determinants of resilience to food security shocks in Southern Somalia*. Mercy Corps, Oregan. Available from: http://bit.ly/1IwOhkK (accessed 11 August 2015).

Millennium Ecosystem Assessment (2005) Millennium Ecosystem Assessment Report. Available from: www.unep.org/maweb/en/Index.aspx (accessed 12 May 2015).

Mitchell T and Maxwell S (2010) *Defining climate compatible development. Climate and development knowledge network policy brief*, London: ODI. Available from: http://r4d.dfid.gov.uk/pdf/outputs/cdkn/cdkn-ccd-digi-master-19nov.pdf (accessed 21 May 2015).

The Montpellier Panel (2012) *Growth with resilience: Opportunities in African agriculture. A Montpellier Panel Report*, London. Available from: http://bit.ly/1Phfyve.

Nasdasdy P (2007) Adaptive co-management and the gospel of resilience. In: Armitage D, Berkes F, and Doubleday N. (eds), *Adaptive co-management: collaboration, learning and multi-level governance*. Canada: UBC Press, 208–27.

Norris F, Betty SP, Pfefferbaum B, Wyche KF and Pfefferbaum RL (2008) Community resilience as a metaphor, theory, set of capacities, and strategy for disaster readiness. *American Journal of Community Psychology*, 41(1–2), 127–50.

Oxfam (2009) People-centred resilience: Working with vulnerable farmers towards climate change adaptation and food security. *Oxfam Policy and Practice: Agriculture, Food and Land*, 9(6), 82–124(43). Available from: http://bit.ly/1HfxF0u (accessed 15 June 2014).

Community and Regional Resilience Initiative, *ResilientUS*: www.resilientus.org.

Reaching Resilience (2013) Reaching Resilience: Handbook Resilience 2.0 for Aid Practitioners and Policymakers. Available from: www.reachingresilience.org/learn.

Rigg J and Oven K (2015) Building liberal resilience? A critical review from developing rural Asia. *Global Environmental Change*, 32, 175–86.

Robards M, Schoon M, Meek C and Engle, NL. (2011) The importance of social drivers in the resilient provision of ecosystem services. *Global Environmental Change*, 21(2), 522–29.

Star SL and Griesemer JR (1989) Institutional ecology, 'Translations' and boundary objects: Amateurs and professionals in Berkeley's Museum of Vertebrate Zoology, 1907–39. *Social Studies of Science*, 19(3), 387–420.

UNDP (2008) *Human Development Report 2007/2008*. New York: UNDP. Available from: http://hdr.undp.org/en/media/HDR_20072008_EN_Complete.pdf.

UNDP (2010) *UNDP Human development report 2010: 20th anniversary edition. The real wealth of nations: Pathways to human development*. New York: UNDP. Available from: http://papers.ssrn.com/sol3/papers.cfm?abstract_id=2294686 (accessed 12 May 2015).

UNDP (2014) *Human development report 2014: Sustaining human progress: Reducing vulnerabilities and building resilience*. New York: UNDP. Available from: http://hdr.undp.org/en/content/human-development-report-2014 (accessed 12 March 2015).

US Indian Ocean Tsunami Warning System Program (2007) *How resilient is your coastal community? A guide for evaluating coastal community resilience to tsunamis and other coastal hazards.* Available from: http://1.usa.gov/1cub7vo (accessed 31 July 2014).

Walker B, Carpenter S, Anderies J, Abel N, Cumming G, Janssen M, Norberg J, Peterson G and Pritchard R. (2002) Resilience management in social-ecological systems: A working hypothesis for a participatory approach. *Ecology and Society*, 6(1), 14.

Walker B, Holling CS, Carpenter SR and Kinzig, A. (2004) Resilience, adaptability and transformability in social–ecological systems. *Ecology and Society*, 9(2), 5.

Watts M (2011) On confluences and divergences. *Dialogues in Human Geography*, 1(1), 84–9.

Wilding, N (2011) *Exploring community resilience in times of rapid change.* Report for Carnegie UK Trust and Fiery Spirits Communities of Practice. Available from www.fieryspirits.com.

World Bank (2008) *Pilot program for climate resilience: Climate funds update.* Washington: World Bank. Available from: http://bit.ly/1li8Uoi (accessed 31 July 2014).

World Bank (2009a) *Making development climate resilient: A World Bank strategy for sub-Saharan Africa.* Washington: World Bank. Available from: http://bit.ly/1RB7K6h.

World Bank (2009b) *World Development Report 2010.* Washington: World Bank. Available from: http://bit.ly/1li8Uoi (accessed 12 May 2015).

World Bank (2013) *World development report 2014: Risk and opportunity: Managing risk for development.* Available from: http://bit.ly/1dgpZtR.

World Food Programme (2010) Conflict then floods: Pakistani family shows resilience. Available from: http://bit.ly/1IzEzwb (accessed 20 October 2013).

WRI, UNDP and World Bank (2008) *World resources: Roots of resilience.* Washington, DC. Available from: http://bit.ly/1RB7K6h.

3 Resilience across disciplines

Having shown how resilience is used in international development debates and policy, and having so far examined framings and discourses around resilience, this chapter looks at scientific knowledge and different perspectives on resilience. It traces the origins of resilience ideas in different fields of science. It examines resilience concepts in ecology, social ecological systems and human development, and its application in disaster risk reduction and community development. It identifies some of the similarities and key differences across these diverse fields. In particular I draw on ideas about resilience from social ecological systems literature and human development to inform international development and environmental change.

Mapping resilience in scientific literature

Chapter 1 introduced definitions of resilience across different fields (see Table 1.2), and this chapter further explores the evolution of resilience ideas in different scientific literature. As already highlighted, resilience has different meanings and has been picked up and applied in policy discourses, in popular media and public debate. This chapter explicitly examines the extent to which multiple strands of resilience scholarship are brought together and some of the core ideas that are shared across fields. It identifies which of these core ideas and concepts are especially relevant – and which do work – for international development and global change.

   This section identifies some of the main scientific fields where resilience concepts are used, building on the preliminary discussion in Chapter 1. Resilience is a term that is used in many different areas of science as well as in more general public discourse. Its use is increasing and also broadening across different scientific fields. A recent analysis by Xu and Marinova (2013) documents the rise in scientific publications on resilience. Their analysis identifies more than 900 cited papers published on the topic of resilience since 1973 (from the Web of Science database) with a strong upward trend. According to Xu and Marinova, and Janssen et al. (2006) the dramatic increase in number of cited publications since 1999 is partly as a result of the establishment of the global Resilience Alliance network and its academic journal *Ecology and Society*, as well as more general increase in scholarship around global environmental change in the 1990s. The analysis is focused primarily on literature on social ecological systems, identifying areas around sustainability, economics, ecology and social science.

Martin-Breen and Anderies (2011) examine a broader range of fields in their comprehensive literature review of resilience across disciplines, and from this they identify three different frameworks for understanding resilience. They term these 'engineering resilience', 'system resilience' and 'complex systems resilience', and they describe them as being of increasing complexity. These interpretations of resilience do not necessarily equate to specific fields or disciplines, although each has its roots in a particular discipline. But they certainly have bearing on the interpretations of resilience in policy discourses (as discussed in Chapter 2), and how resilience is understood by different audiences.

What Martin-Breen and Anderies refer to as *engineering resilience* concerns the ability to withstand a large disturbance without in the end changing or becoming permanently damaged – it is about *bouncing back*. It equates to many popular understandings of the term. It is termed engineering resilience because it is conventionally used to refer to how engineered systems, such as bridges, buildings and other infrastructure are designed to handle large stresses and return quickly to normal once the stress is removed. Martin-Breen and Anderies show how elements of engineering resilience are found in earlier understandings of resilience in child developmental psychology (as discussed later in this chapter) and responses to disasters. It also resonates strongly with articulations of resilience in the policy discourses presented in Chapter 2.

According to Martin-Breen and Anderies, *system resilience* recognises a world in flux, where there is no fixed 'normal', but that there are fixed functions within social, technological, ecological and perhaps economic systems which are necessary to continue functioning. Resilience here is about *maintaining system function* in the event of a disturbance. So the system – however defined – is seen as dynamic but remains functioning in the face of different types of change, including long-term slow changes that are more difficult to recognise and measure.

However, new systems may be created in response to disturbance; in other words –using terminology from social ecological systems – they might self-organise and show adaptive capacity. For Martin-Breen and Anderies it is this adaptability that distinguishes a complex adaptive system. Resilience here is defined as the 'ability to withstand, recover from and reorganise in response to crises' (p2); function is maintained, but system structure may change. Transformability may also be a feature, representing the ability of a complex adaptive system to assume a new function. These distinctions are important because transformative resilience becomes a theme I develop in this book, as I explore the implications and applications for international development theory, policy and practice. For example, as discussed in the previous chapter, current responses to climate change emphasise incremental adaptations, often 'climate proofing' current activities. Whereas more radical approaches (which underpin transformational system change) would integrate risk and uncertainty in the design of new institutions, new forms of food production and energy provision which contribute to both adaptation and mitigation of climate change.

How much overlap is there between the different scientific fields that engage with and use resilience concepts? In other words, how distinct are the fields? Jacopo Baggio, Denis Hellebrandt and I undertook a bibliometric analysis to identify where resilience ideas are used in scientific literature (Baggio et al. 2015). We analysed 994 published papers which had the term 'resilience' in their title and, using tools from network analysis, we examined the extent to which different fields and sub-fields were connected through citations. We identified five separate fields: social sciences (including economics); ecology and environmental sciences; psychology; engineering; and social ecological systems, each representing different disciplinary perspectives. Despite the claims in some literature that resilience is a boundary object or bridging concept (a term with a precise meaning within fields that is also used loosely across fields, or purposely to integrate different fields), we found there was limited citation and cross-referencing between these fields. Most papers prefer to refer to and cite exclusively within their own specific field, if not sub-field. The most interdisciplinary papers are within the social ecological systems literature where the majority of papers are cited at least 50 per cent of the time outside of their own field. Psychology, social sciences and ecology are more 'exclusive' in the use of the concept of resilience, and have fewer connections with other fields.

Whilst our analysis confirms the rise of the term resilience and its use across different fields, we show that resilience does not seem to bridge different fields, at least in the academic literature. The engineering field is the most isolated, with very few citations to and from other fields. Some sub-fields in the social sciences seem to be almost, if not completely, isolated (e.g. industrial organisation and business). Ecology and psychology interact in the main only within their own field. The social ecological systems literature has most connections with the ecology literature. As might be expected, and as shown in other studies (Xu & Marinova, 2013; Janssen et al. 2006) Holling's 1973 paper that introduced the concept of resilience in ecology is the most important central paper. Social ecological systems literature, perhaps because of its interdisciplinary origin, interacts more than all other fields, and it cites literature from the social sciences, engineering, psychology and from ecology especially. Overall there is relatively little integration between social ecological systems and human development – despite shared concepts and similar developments and concepts, other than a few key papers.

Discussion now turns to focus in more detail on three areas of scientific inquiry where there has been almost parallel development of resilience ideas.

These speak much more directly to my own understanding, interpretation and application of resilience. I explore the evolution of resilience concepts in two quite distinct fields – in social ecological systems, and in human development. I then discuss how these are applied in analysis of disasters (specifically in the sub-field 'disaster risk reduction') and social and community resilience, as key areas informing international development and environmental change.

Resilience in social ecological systems

Carl Folke, founding Director of the Stockholm Resilience Centre and co-editor of the journal *Ecology and Society*, has written a definitive account of the emergence of resilience concepts in ecology and its evolution to a wider, interdisciplinary systems perspective in terms of social ecological systems (Folke 2006). In this account, Folke maintains that the resilience approach has its roots in one branch of ecology: the discovery of 'multiple basins of attraction' (p254) in ecosystems in the 1960s and 1970s and hence 'inspired social and environmental scientists to challenge the dominant stable equilibrium view' (Folke 2006: 253). This stable equilibrium view understood ecosystems to evolve – by the process of succession – towards a stable state or climax where they cease to grow or change. The alternative ('multiple basins of attraction' – see definition in Box 3.2) postulated that ecosystems were in a more dynamic flux, displaying multiple equilibria or alternate states.

Resilience ideas in ecology emerged from studies of interacting populations of predators and prey and their functional responses in relation to ecological stability theory. The seminal paper by C.S. (Buzz) Holling published in 1973 on resilience and stability in ecological systems explored the existence of 'multiple stability domains' or 'multiple basins of attractions' in natural systems and how they relate to ecological processes, random events such as disturbance, and different temporal and spatial scales. In this paper resilience was defined as determining the persistence of relationships within a system and as a measure of the ability of these systems to absorb changes and still persist. This approach, recognising multiple stable states rather than a single equilibrium, represented a very different way of understanding ecosystems and offered new directions and analyses, involving non-linear relationships, uncertainty and shifting or multiple boundaries. Whilst Holling's ideas had, in his own words (Folke 2006: 254) 'excellent causal and process evidence, the field evidence was only suggestive'. However, there were significant and far-reaching implications for ecosystem management as well as theory.

Through the late 1970s and 1980s ecologists started applying these ideas to field-based studies, producing some of the iconic case studies that are widely cited in the literature. These include studies of boreal forest dynamics (Holling 1978; Ludwig et al. 1978) and rangelands (Walker et al. 1981). Box 3.1 highlights some of the iconic examples of resilience studies.

Box 3.1 Iconic studies of social ecological systems resilience

The Everglades, Florida, USA: a world famous wetland system, where significant parts of the protected area, due to increased eutrophication and changes in water levels and shocks such as drought and fire, have crossed a threshold into a system dominated by 'cattail' grasses.

The Goulburn-Broken Catchment in Australia: increasingly intensive agriculture is undermining system resilience, and saline groundwater is perilously close to the surface.

Coral Reefs in the Caribbean: multiple stressors, including over-fishing, have resulted in loss of 80% of hard corals and perhaps resulted in irreversible changes to the ecosystem.

Northern Highland Lakes District in Wisconsin, USA: an area of 7,500 lakes which are at risk from a variety of pressures, including pollution, climate change, invasive species, over-harvesting and deforestation.

Kristianstad Water Vattenrike, Sweden: an internationally renowned wetland where biodiversity and habitats are disappearing, as a result of increasing water control measures.

Source: Adapted from Walker and Salt 2006

A number of core concepts developed from ecological studies of resilience have been applied to more interdisciplinary analysis of social ecological systems. I provide definitions of some of the most often used in Box 3.2, and then a brief explanation in this section of core concepts and heuristics (also see the Glossary). The concepts are important to draw parallels with ideas in other fields and because they underpin the dynamic systems thinking which is central to resilience approaches, and illuminate applications of resilience in different contexts.

The adaptive cycle

The ideas and logic behind the adaptive cycle underpin many of the assumptions and analysis of dynamics of change in ecological and social ecological systems. The model of the adaptive cycle was derived from the comparative study of the dynamics of ecosystems and was first suggested by Holling in 1973. The adaptive cycle is described as a *heuristic device* – in other words as an artificial construct to assist in the exploration of

Box 3.2 Core concepts in social ecological systems resilience

Adaptive cycle is a way of describing the progression of a social ecological system through various stages of organisation and function.

Basin of attraction is a metaphor to help visualise alternative state regimens, where a social ecological system can exist in one or more system configurations.

Controlling variables (e.g. nutrient levels in a lake, depth of the water table) determine the levels of other variables (like algal density of soil fertility).

Slow variables are controlling variables which change slowly (sediment concentrations, population age structures) and which determine the dynamics of the fast variables which are of direct interest to managers.

Fast variables are often those on which the human use of the system is based.

State variables define the system – for example a rangeland system might be defined by the amounts of grass, shrubs and livestock. The set of state variables together with the interactions between them and the processes and mechanisms that govern them define the system.

A **threshold** occurs where the levels of the underlying controlling variables change and this feeds back to the state of the system.

Panarchy is a hierarchical set of adaptive cycles at different scales in a social ecological system and their cross-scale effects.

Social ecological system is a complex integrated system of people and nature.

Source: Walker and Salt 2006, Walker et al. 2004

phenomena. It involves assumptions derived from empirical research, although it can also have explanatory value as a model. It focuses attention on processes of destruction and re-organisation, which are often neglected in favour of growth and conservation. Including these processes, it is claimed, provides a more complete view of system dynamics that links together how the system itself is organised, how it responds to disturbance and how it changes.

The Resilience Alliance[1] explains the adaptive cycle as a fundamental means of understanding complex systems, ranging from cells, to ecosystems, and even to societies. An adaptive cycle alternates between relatively long periods of aggregation and transformation of resources, and shorter phases that create opportunities for innovation (see also Walker & Salt 2006).

For ecosystem and social ecological system dynamics that can be represented by an adaptive cycle, four distinct phases, shown in Figure 3.1, are identified:

- Rapid growth or exploitation (r) – a phase in which resources are readily available, and entrepreneurial agents exploit niches and opportunities.
- Conservation (K) – resources become increasingly locked up and the system less flexible and responsive.
- Release (Ω) – a disturbance causes a chaotic unravelling and release of resources.
- Re-organisation (α) – new actors (species, groups) and new ideas can take hold; generally leads to another r phase.

Thus the adaptive cycle has two major phases. The first, from r to K – referred to as the foreloop in social ecological resilience literature – is a slow period of incremental growth and accumulation. The second phase – omega to alpha – of rapid re-organisation, that leads to renewal. This is the backloop.

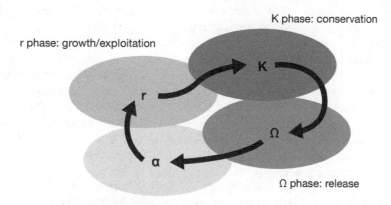

Figure 3.1 The adaptive cycle

Source: Figure 2–1 (page 34) from Panarchy edited by Lance H. Gunderson and C.S. Holling. Copyright © 2002 Island Press. Reproduced by permission of Island Press, Washington, DC

During exploitation and conservation, connectedness and stability increase and nutrients and biomass (in ecosystems) slowly accumulate. This 'capital' is built up and stored in the system. In an ecosystem, competitive processes usually lead to a few species becoming dominant, with diversity retained in residual pockets in a patchy landscape. The capital is stored for the growing, maturing ecosystem, and it also represents a gradual increase in the potential or options for other kinds of ecosystems in the future. In an economic or social system, the accumulating potential could be from the skills, networks of human relationships, and mutual trust that are incrementally developed during the progression from r to K.

Opportunities for innovation appear at different points in the cycle of social ecological disturbance or renewal. Social ecological systems often go through cycles of growth and development – for example, ecosystem succession, development of management expertise in an agency, then a conservation phase where conditions are relatively stable, then perhaps a radical change (release) due to a shift in ecological or political environment, then a re-organisation (Gunderson & Holling 2002). The strategies stimulating innovation and novelty differ among these phases of the cycle of disturbance and renewal.

This conceptualisation departs from traditional ecological understanding that focused on the notion of succession, which describes the transition from exploitation (i.e., the rapid colonisation of recently disturbed areas) to conservation (i.e., the slow accumulation and storage of energy and material). Two additional functions – release and re-organisation – are necessary to describe the system dynamics in the adaptive cycle.

Panarchy

As explained by Walker and Salt (2006), social ecological systems have structures and functions that cover a wide range of spatial and temporal scales. Structures and processes are linked across scales, and these interactions can occur both bottom up and top down. This means that we cannot understand the dynamics of change at any one scale without considering what happens at other scales. For example, an assemblage of individual plants makes up a woodland or forest system, but each plant or each tree has a set of insects, fungi and bird species associated with it, each has its own dynamic, but each is linked. *Panarchy* is a cross-scale, nested set of adaptive cycles, shown in Figure 3.2.

The figure in Gunderson and Holling's edited collection *Panarchy*, (2002: 75) (reproduced here as Figure 3.2) illustrates three levels of panarchy and shows the two connections that are critical in creating and sustaining adaptive capability. One is the 'revolt' connection – this can cause a critical change in one cycle to cascade up to a vulnerable stage in a larger and slower one. The other is the 'remember' connection, which facilitates renewal by drawing on the potential accumulated and stored in the slower, larger cycle.

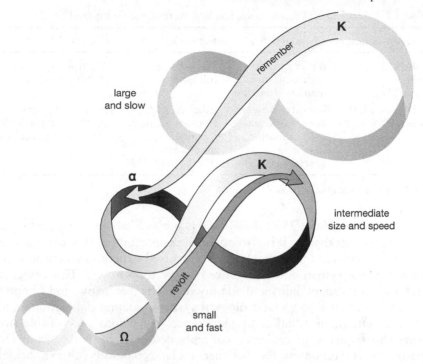

Figure 3.2 Panarchy

Source: Figure 2–1 (page 34) from Panarchy edited by Lance H. Gunderson and C.S. Holling. Copyright © 2002 Island Press. Reproduced by permission of Island Press, Washington, DC

Social ecological systems

As scholarship on resilience grew, so the applications extended beyond ecosystems to consider a social ecological system, as human interactions and management of ecosystems became integrated into analysis of dynamics of change. Folke (2006) charts the evolution of resilience thinking from ecology to broader social ecological systems. Walker and Salt (2006: 164) define social ecological systems simply as 'linked systems of people and nature'. Berkes, Folke and Colding (1998) used the term 'social-ecological system' to emphasise the integrated concept of humans-in-nature, to emphasise the delineations between humans and nature as arbitrary and artificial. They define social ecological systems as 'complex, integrated systems in which humans are part of nature'. The Resilience Alliance glossary[2] expands on this, explaining that, 'evidence suggests that social-ecological systems act as strongly coupled, complex and evolving integrated systems'. Social ecological systems encompass a wide range of different resource management regimes, institutions across different spatial and social scales.

Work on ecological or ecosystem resilience emphasised the capacity to absorb disturbance, or to build the capacity to buffer that allows persistence.

Table 3.1 Ecological and social ecological system resilience compared

Resilience concepts	Characteristics	Focus on	Context
Ecological resilience	Buffer capacity, withstand shock, maintain function	Persistence, robustness	Multiple equilibria, stability landscapes
Social ecological resilience	Interplay between disturbance and re-organisation, sustaining and developing	Adaptive capacity transformability, learning, innovation	Integrated system feedback, cross-scale dynamic interactions

Source: Adapted from Folke 2006

But Folke notes (2006: 259): 'Resilience is not only about being persistent or robust to disturbance. It is also about the opportunities that disturbance opens up in terms of the recombination of evolved structures and processes, renewal of the system and emergence of new trajectories.' This brings a greater focus on adaptability and adaptive capacity (see below, and Chapter 5). So when applied to social ecological systems, adaptation, learning and self-organisation, in addition to persistence, come into play. Table 3.1 shows this sequence of resilience concepts described by Folke, from the narrow interpretation to the broader social ecological context. Social ecological resilience is then understood (following Carpenter et al. 2001 and presented by Folke 2006: 259–60) as comprising three aspects:

- the amount of disturbance a system can absorb and still remain within the same state;
- the degree to which the system is capable of self-organisation;
- the degree to which the system can build and increase the capacity for learning and adaptation.

Social ecological resilience forms a central plank for much of the discussion about resilience in this book; it is *one* of my starting points.

Social resilience

So to what extent does the evolution in resilience concepts to a social ecological systems perspective integrate understanding of social dimensions, and how does this relate to social resilience? I have discussed the 'social turn' in resilience thinking (Brown 2014) which highlights three topics around which recent interest in social dimensions of resilience are taking shape and advancing: in applications of community resilience; societal transformation; and social transitions. I discuss later in this chapter the applications of these ideas around community resilience, which uses concepts from social ecological systems and psychology fields particularly. But the social dimensions of resilience within the social ecological systems approach are not unified into any single voice, nor do they adhere to one definition or analytical or conceptual framework. Different aspects of the 'social' are emphasised and are analysed in different studies. Most conceptual frameworks of a social ecological system are represented by a diagram which show two sub-systems, the social and the ecological, linked perhaps, within a larger social ecological system which is then influenced by a set of external stresses. The aim is to develop an understanding that provides a stronger conceptual linkage between the social and the ecological.

Neil Adger's work is frequently cited in this field; in his 2000 paper he defines social resilience as 'the ability of communities to withstand external shocks to their social infrastructure' (p361). Ecological and social resilience are linked, and he finds parallels between them. Adger identifies resource dependence and institutions as two key elements in this linkage. This determines how tightly coupled the social and ecological elements are, and how responsive they each are to changes in one or the other. Certainly the social ecological systems literature has put much emphasis on institutions as key mediating factors, linking the social and the ecological in a two-way relationship influencing resilience.

My approach picks up threads of research on social aspects of resilience from different literatures to develop a more integrated view of resilience, building on social ecological systems approaches but bringing in ideas from human development, community resilience and disasters, and from research on adaptation and adaptive capacity. One useful definition of social resilience comes from Peter Hall and Michele Lamont, writing about the impacts of neoliberalism, who understand social resilience as multi-scalar and dynamic and as an essential characteristic of 'successful societies'; as 'the capacity of groups of people bound together in an organisation, class, racial group, community, or nation to sustain and advance their wellbeing in the face of challenges to it' (2013: 2). This emphasises the collective nature of social resilience, but extends beyond the often place-based analysis of people bound to a particular resource; and also relates the definition to their well-being. This also resonates clearly to the human development approaches examined in the next section.

The broadening of resilience concepts from ecology or ecosystems, to encompass a much wider social ecological systems perspective represents a shift not only in focus and from single to multiple and interdisciplinary analysis, it also integrates a new set of ideas around adaptation and adaptive capacity, learning and innovation. I discuss these in Chapter 5 particularly. This wider perspective interfaces more directly with other disciplinary approaches to understanding how individuals and systems deal with change. It also enables resilience concepts to more cogently address and inform sustainability agendas. I discuss these applications later in this chapter, but the next section looks at the almost parallel evolution of resilience ideas in the field of human development.

Resilience in human development

Resilience is a term widely applied in the field of child and developmental psychology, and in related fields of anthropology, sociology and psychology. I use insights from these fields alongside those from social ecological systems to develop a revised and socially informed perspective on resilience. I use the term human development to cover these literatures, and to refer to anthropological, sociological, psychological and biological approaches to studying human development. In this literature, individual resilience is defined as 'the dynamic process wherein individuals display positive adaptation despite experiences of adversity or trauma' (Luthar & Cicchetti, 2000: 81). Resilience here has three dimensions: positive outcome despite the experience of adversity; continued positive or effective functioning in adverse circumstances; and recovery after significant trauma. These strongly resonate with the dimensions of resilience – coping, bouncing back and adapting – introduced earlier in Chapter 1. A fourth dimension has recently been posited, where resilience also includes positive transformation following adversity, which relates to a positive re-organisation of systems with adaptive functioning being better than previously. This evolution in the conceptualisation of resilience resonates with the distinctions made by Martin-Breen and Anderies (2011) discussed at start of this chapter. In some respects, it mirrors developments in other fields of resilience, such as the broadening from ecological to social ecological systems.

Resilience research in human development emerged in the 1960s and 1970s when the phenomenon of positive adaptation was observed in subgroups of children considered 'at risk' for developing later psychiatric disorders (Masten 2007). Thus at a similar time that resilience ideas were developing in ecology, resilience researchers in developmental psychology were trying to understand, prevent and treat mental health problems, such as schizophrenia and autism, and to understand the consequences of naturally occurring stressors, such as death, trauma, poverty and prenatal risk in the family (Luthar 2006).

Since the 1960s, the study of resilience has advanced in four waves of research (Luthar 2006; Masten 2007). In the first wave, resilience was considered a stable personal characteristic and research focused on identifying various individual, family and community protective factors associated with positive adjustment despite exposure to risk factors. The second wave led to a more dynamic account of resilience, with researchers adopting a developmental-systems approach to theory, focusing on the transactions – or interactions – among individuals and the different processes and mechanisms through which exposure to risk factors may be associated with children's positive and negative outcomes. The third wave focused on creating resilience by preventive interventions directed at changing developmental pathways (i.e. specific protective processes), promoting competence and wellness. The fourth wave builds on work in neurobiology,

Table 3.2 Four waves of resilience research in human development

Wave	Year, Key reference	Focus	Shift in focus of resilience
1	1999 Masten	Identifying the correlates of resilience in individual children and families, drawn from studies of individuals. Resilience linked to personal qualities or traits.	Characteristics of individuals
2	2006 Schoon	Characterising the processes which might account for the correlates, using longitudinal studies and focusing on relationships and interactions.	Relationships
3	2006 Luthar	A focus on prevention and on formulating interventions to foster resilience, including experiments	Influence of society
4	2007 Masten	Multi-level system dynamics	Complex systems

Source: Masten et al. 2013

genetics and development to enable risk and resilience processes to be studied from multiple levels of analysis and their interplay (Masten & Obradović 2008; Masten 2007).

These four waves are significant, because they trace the development of ideas and concepts surrounding resilience in these fields, and they relate to current debates and to the evolution of ideas across fields. The waves represent a shift in focus from individual characteristics of resilience to more complex interactions of society and family, and more dynamic, multi-scalar analyses. Table 3.2 summarises these 'waves' of research, and the following sections explain them in more detail.

First wave: identifying protective factors

Resilience is attributed to individuals who beat the odds, bounce back or avoid the negative trajectories associated with risks even though they have been exposed to risk factors. Risk or adversity can comprise biological, psychological, genetic, environmental or socio-economic factors that are associated with an increased probability of 'maladjustment' (Masten et al. 1990). Many kinds of adversity and their impact on individual adjustment have been investigated, such as premature birth, mental illness in a parent, divorce, family violence, community violence or war. All these factors are associated with negative outcomes in children, but they are not necessarily equivalent in severity; rather, severity depends upon both the risk factor and the population in question. However, a major risk factor influencing individual adjustment that affects children's cognitive competences as well as social-emotional functioning is socio-economic adversity (Luthar 2006; Schoon 2006). This encompasses living conditions characterised by low social status, poor housing, overcrowding, poor quality schools and lack of material resources.

In addition to single risk factors (such as community violence, poverty and parental mental illness) or experience of a stressful life event (for example, parental divorce and bereavement), composites of multiple risk indices, such as parents' low income and education, histories of mental illness and disorganisation in children's neighbourhoods have been studied (Luthar 2006). Importantly, Rutter (1979) showed that when risks coexist, effects tend to be *synergistic*, with children's outcomes being far poorer than when any of these risks existed in isolation. Hence, high-risk individuals have life histories characterised by multiple disadvantages and an accumulation of risk effects, and children living in extreme or chronic poverty often have the worst outcomes (Boyden & Cooper 2007). This resonates strongly with work on vulnerability in environmental change literature, which refers to ratchet effects of poverty, and the synergistic effects of multiple stressors, for example by Leichenko and O'Brien (2008). I explore these aspects in terms of lived experiences of resilience in Chapter 4 and in the context of poverty in Chapter 6.

Yet in this wide literature on human development, there is no consensus about what constitutes adaptive functioning, and definitions of successful adaptation differ in regard to historical, cultural and developmental contexts. A focus on positive outcomes may consider the maintenance or return to adequate functioning after the experience of adversity or trauma, in addition to trying to understand how individuals develop new coping skills (Kaplan 1999). In terms of positive adjustment, some studies look at the absence of psychological disorders, while others require more positive outcomes such as academic achievement, social competence or meeting stage-salient developmental tasks (Masten 2001; Ungar 2004). Debate continues over whether positive adjustment should be used for exceptional attainments or for more ordinary achievements – in other words, for bouncing back or bouncing forward.

The first wave of resilience research identified resilient individuals who were 'doing okay' compared to others who were not adapting when faced with similar risks or adversities. This work looked at linkages among characteristics of individuals and their environments that lead to a good outcome in face of high risk or adversity. A 'short list' of broad correlates of better adaptation among children at risk was developed (see Box 3.3), identifying a set of individual child, family, community and societal characteristics which represent protective factors or assets contributing to resilience. Masten (2001) observes that these show remarkable consistency and ultimately constitute or reflect the fundamental adaptive systems supporting human development. She considers that in the majority of cases, resilience results from ordinary adaptive processes rather than extraordinary ones, which she refers to as 'ordinary magic'.

Box 3.3 Assets and protective factors

Child characteristics	Family characteristics
• social and adaptable temperament in infancy • good good cognitive abilities and problem-solving skills • effective emotional and behavioural regulation strategies • positive view of self (self-confidence, high self-esteem, self-efficacy) • positive outlook on life (hopefulness) • faith and a sense of meaning in life • characteristics valued by society and self (talents, sense of humour, attractiveness to others).	• stable and supportive home environment • parents involved in child's education • parents have individual qualities equivalent to the child characteristics in upper left quadrant • socio-economic advantages • postsecondary education of parent • faith and religious affiliations.
Community characteristics	**Cultural or societal characteristics**
• high neighbourhood quality • effective schools • employment opportunities for parents and teens • good public health care • access to emergency services (police, fire, medical) • connections to caring adult mentors and pro-social peers.	• protective child policies (child labour, child health, and welfare) • value and resources directed at education • prevention of and protection from oppression or political violence • low acceptance of physical violence.

Source: Adapted from O'Dougherty Wright and Masten, 2005

Second wave: resilience in developmental and ecological systems

The second wave of resilience research reflects a broader view of normative and pathological development. Resilience research 'has increasingly focused on contextual issues and more dynamic models of change, explicitly recognising the role of developmental systems in causal explanations' (O'Dougherty Wright & Masten 2005: 25). The role of relationships and systems beyond the family has become emphasised with biological, social, historical and cultural processes being integrated into models and studies of resilience (e.g. Masten 2001; Schoon 2006). Hence, resilience research is more contextualised such as how the individual interacts with many systems at different levels throughout their life course, and is careful of making generalisations about risk and protective factors between contexts or developmental stages (O'Dougherty Wright & Masten 2005).

In this second wave of resilience thinking in human development, the influence of developmental systems theory is evident in the dynamic, reciprocal, multi-causal and multiple level models of resilience, developmental pathways and trajectories. Studies have explored the moderating processes that explain protective effects that work for some people in some conditions, and the mediating processes that explain how risk or protection work to undermine or enhance adaptation (O'Dougherty Wright & Masten 2005). In many ways this mirrors the evolution of resilience thinking in complex adaptive systems, and the move from ecology to social ecological systems.

This marks a shift away from a focus on the individual to a broader focus, which encompasses family and community relational networks in order to understand the process of positive or negative adaptation following stress (Masten 2007). Studies also provided a more complex assessment of family and environmental influences (Rutter 2000). Since children respond to adversity in different ways and there are variations in adjustment (Luthar 2006; Schoon 2006), a 'transactional model of influence' provides an important opportunity to focus on transitions or turning points in individual's lives (Masten & Reed 2002). Studies also show that there are critical turning points in response to specific developmental challenges, such as entering school or undergoing adolescence, that can shape the nature and course of future adaptation. Once again this resonates with ideas about thresholds and transitions that have emerged from the study of environmental change and social ecological systems.

However, authors point out that it is damaging to view resilience as an individual trait; when children do not adapt successfully they could be seen as being personally to blame for not being able to cope with adversity (Schoon 2006). Often, the same forces that can constrain the child's development (poverty, discrimination, inadequate medical care or exposure to community violence) impact and constrain the entire family. In these circumstances, parents are unable to provide the necessary resources and basic protection their children need, particularly as they lack support from the wider society. Hence Ungar (2005a) reports that resilience is as much dependent on the structural conditions, relationships and access to social justice that children experience as it is on any individual capacities. This echoes debates in social science literature that point to potential for resilience approaches to over-emphasise the role of individual and under-emphasise the structural elements that undermine resilience and bring about vulnerability (Brown 2014). The cases discussed in Chapter 4 highlight interactions between the personal or individual, and the external system or structural aspects that affect resilience in different contexts.

The movement away from an individually based conceptualisation of resilience towards a contextually-situated framework – the shift from first to second wave – recognises the importance of hitherto poorly understood cultural factors. Many other factors have been identified within the collective network of the family and the community. For example, for various cultural or ethnic groups there can be a great deal of difference in the importance placed on individualism, collectivism and the family, and these dimensions might affect resilience in different ways for different groups. Similarly how particular cultural groups define and manage risk may be expected to vary in accordance with whether they attribute outcomes to fate, supernatural forces or human agency. There are insights for international development, and the relationship between work on poverty and resilience, and for understanding socially differentiated adaptive capacity. Camfield and McGregor (2005) explore well-being and resilience in developing countries in order to integrate both objective and subjective approaches to understanding how poverty is experienced and reproduced. Brown & Westaway (2011) highlight the relevance of agency in relating poverty and resilience. These themes are further discussed in Chapters 4, 5 and 6.

Third wave: interventions to foster resilience

A major reason for studying resilience is to inform practice and policy to create resilience in situations where it is lacking or absent. The third wave of resilience research in human development focused initially on theory-driven intervention designs and then moved on to experimental studies to test resilience theory (O'Dougherty Wright & Masten 2005). Many of the studies showed that promoting competence was a key element of programmes that worked, with mediators and moderators of change being similar to the processes implicated by the 'short list' in resilience research (see Box 3.3). This relates to current policy focus in 'building resilience' in response to environmental (and specifically climate) change, and in applications in disaster risk reduction.

In the early 2000s, a 'resilience framework' for practice and policy was described (Luthar & Cicchetti 2000; Masten 2001), and intervention models are described as a protective process to promote resilience. In recent years, resilience research has advanced many models of positive development and strength-based models (Masten 2007), which acknowledge children as competent social agents and capture their subjective experiences (Boyden & Cooper 2007). Schoon (2006) presents a developmental-contextual systems model for the empirical study of adaptations in context and time, which emphases multiple interacting spheres of influence and life course theory.

The model advocates a systems view of resilience, postulating a holistic approach incorporating multi-level person-context interactions. The model also captures the interacting nature of development over time, focusing on the reciprocal interactions between risk experiences and individual adjustment, which are embedded in the wider socio-historical context; and transitions. Schoon's model is shown in Figure 3.3. The striking thing about the diagram, and the way it represents cyclic relationships of interactions and feedbacks linked across scales, is the way it resembles ideas about panarchy and even the representation of panarchy in Figure 3.2.

Analyses of preventive programmes that work for children underscore the importance of programmes that embrace a developmental, 'ecological' systems approach (O'Dougherty Wright & Masten 2005). *Ecological* here refers to a systems-based understanding which recognises multiple interacting factors at different scales. These comprehensive prevention approaches acknowledge the multiplicity of risks and cumulative trauma faced by many children in different circumstances and at different life stages, and emphasise the importance of promoting competence and well-being, and building protection across multiple domains in order to achieve a positive outcome (Schoon 2006). Hence, interventions aim for a holistic approach, focusing on individuals but working with communities and families, and identify windows of opportunity for intervention at key transition points (Masten 2007).

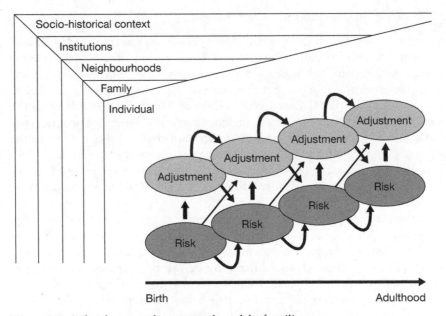

Figure 3.3 A developmental-contextual model of resilience

Source: Brown and Westaway 2011. Originally adapted from Schoon 2006

But a critical challenge for third wave researchers has been to address discrepancies between research findings (on the nature of risk, protective factors, resilience and recovery) and public policy. This means that not only do individual interventions matter, but also system and structural change: 'Ultimately, for interventions to be successful they should lead to a complete and permanent ecological change, reinforcing diversions in development in the long run through on-going structural improvements' (Schoon 2006: 169). Interventions need to address causal systemic and structural issues, not just target individuals and specific risk factors. Again, this has relevance for discussions about the application of resilience in other fields, including planning and disaster recovery, and for international development and environmental change.

Fourth wave: multi-level dynamics

As a systems perspective on resilience in human development fields has grown stronger, attention has shifted to multi-level dynamics or the ways in which resilience is shaped by interactions across levels of analysis. These include gene-environment interactions; brain function; social interactions and co-regulation among individuals in relationships and social networks; and person-media interactions. Masten (2007) believes this fourth wave of resilience research will overtake and assimilate earlier work; being fed by new technologies and the synergy from integrative theory and methodology. The fourth wave focuses on integrative studies across multiple levels of analysis. Thus, the study of risk, assets, vulnerabilities and protection mapping at multiple levels using statistical tools to model complex dynamics has enabled the more in-depth study of processes of resilience in human development. In reconceptualising resilience, Luthar (2006) considers that resilience rests fundamentally on relationships – strong, proximal relationships providing the warmth, support and informal social control that serve protective processes which are critical for achieving and sustaining resilient adaptation for children, as well as for adults.

A constructivist approach to resilience

But just as in the fields of social ecological systems, geography and environmental change, there are multiple views and contestations around resilience in human development. An important critical voice is that of Michael Ungar, co-director of the Resilience Research Centre at Dalhousie University in Halifax, Canada.[3] He challenges what he refers to as the dominant ecological understanding of resilience in human development which is underpinned and informed by systems theories and which emphasise the relationships between risk and protective factors, circular causality and transactional processes that foster resilience (Ungar 2004). Ungar proposes an alternative constructivist discourse on resilience, where, instead of resilience being an objective fact, he defines resilience as 'the outcome from negotiations between individuals and their environments for the resources that define themselves as healthy amidst conditions collectively viewed as adverse' (Ungar 2004: 342). With this a more qualitative approach to researching resilience is advocated, and a challenge for researchers and practitioners to examine how issues such as race, gender, class, ability and other factors affect not only access to health resources but at a more fundamental level, the definition of resilience itself (Ungar 2004). These are critical issues when considering how resilience ideas might inform international development.

Ungar applies this approach to the analysis of children's narratives of resilience across 14 different contexts as part of the International Resilience Project (see Ungar et al. 2007). This study aims to provide a more contextually relevant understanding of resilience. It identifies seven factors – or 'tensions' – that influence resilience for young people. The seven factors (shown below in Table 3.3) are found in each of the cultural contexts studied, but each plays out differently, according to the individual, their family and community, and the culture in which they live. The authors present the tensions as a 'conceptual map' (p295) to help explain what resilience means for young people across diverse cultures and circumstances. Importantly they find that youth who experience themselves to be resilient and are seen by their communities as resilient are able to successfully navigate their way through these tensions (Ungar et al. 2007). They found that the tensions themselves are dynamic – they converge in different ways across time, and throughout a person's life-course, and what is most important is the way that they interact. It is at the intersections of the seven tensions that what constitutes resilience in any given culture or context is revealed. There is no optimal way of dealing with or reconciling these tensions; rather, Ungar et al. suggest resilience is about finding a way to live in relative comfort despite contradictions and conflicts; to continue to navigate and negotiate challenges. In this way, resilience is not a permanent state of being, but a condition of becoming better.

This application of a more qualitative, constructivist approach, resonates strongly with some of the social criticisms of resilience in environmental change and social ecological system literature, and how it uses similar terminology to recent studies in social ecological systems literature. So it uses expressions like 'navigating change', 'turbulence', 'identity', 'power' and 'control'. It provides a helpful bridge between these fields, and suggests how resilience can be re-articulated to focus on capacities and strengths, recognising humans in different cultural contexts. In emphasising resilience as socially constructed and providing cross-cultural perspectives it acknowledges the importance of narratives – which I discuss in the next chapter – and resilience as an on-going process of negotiation and navigation, of interaction and dynamic tensions, and between resilience as means and ends. The tensions themselves are also similar to emerging themes in research on adaptive capacity in Chapter 5.

Table 3.3 Seven tensions influencing resilience in young people

Tension	Explanation
1 Access to material resources	Availability of financial, educational, medical and employment assistance and/or opportunities, as well as access to food, clothing and shelter
2 Relationships	Relationships with significant others, peers and adults within one's family and community
3 Identity	Personal and collective sense of purpose, self-appraisal of strengths and weaknesses, aspirations, beliefs and values, including spiritual and religious identification
4 Power and control	Experiences of caring for one's self and others; the ability to affect change in one's social and physical environment in order to access health resources
5 Cultural adherence	Adherence to one's local and/or global cultural practices, values and beliefs
6 Social justice	Experiences related to finding a meaningful role in community and social equality
7 Cohesion	Balancing one's personal interests with a sense of personal responsibility to the greater good; feeling a part of something larger than one's self socially and spiritually

Source: Ungar et al. 2007: 295

Resilience applications

Having given an overview of two major fields of resilience scholarship, I now focus on some of the sub-fields or applications of resilience ideas. I look at how resilience concepts are used in writing around disasters and natural hazards, and social and community resilience related to environmental change. These areas are where the two fields – social ecological systems and human development – have, to an extent, been integrated. They are also closely allied to, and inform international development, partly because of development agencies' role in humanitarian aid and disaster relief – now more often framed as disaster risk reduction.

The terms 'resilience', 'risk' and 'vulnerability' are used prominently in the analysis of hazards and disasters. In this literature, there is much discussion on the evolution, convergence and continuing distinctions between different approaches, their semantics, meanings and practices. For some, the terms are almost interchangeable, with resilience being an antonym of vulnerability, whereas for others, a move to resilience terminology represents a paradigmatic shift in thinking and policy and practice. There is much debate especially in the disaster risk reduction literature about the significance of this move towards a resilience approach. In their review of resilience ideas and their applications to climate change adaptation and disaster risk reduction, Bahadur et al. (2010) refer to a 'Resilience Renaissance' (p2). They review 16 different conceptualisations of resilience and distil a set of common factors identified as important components of resilience. The ten characteristics of resilient systems are: high diversity; effective governance, institutions or control mechanisms; acceptance of uncertainty and change; community involvement and inclusion of local knowledge; preparedness, planning and readiness; high degree of equity; social values and structures; non-equilibrium system dynamics; learning; and adoption of cross-scalar perspective.

A number of authors view vulnerability simply as the antonym of resilience. Others view vulnerability as quite distinct from resilience. For example, Gallopin (2006) contends that vulnerability is the capacity to preserve the structure of a system, whereas resilience refers to its capacity to recover from non-structural changes in dynamics. Nelson et al. (2007) contend that a resilience approach differs from a vulnerability approach in a number of key ways shown in Table 3.4. This distinction is further discussed in Chapter 5.

An important strand in the disasters literature focuses on community resilience. Cutter's Disaster Resilience of Place (DROP) model (Cutter et al. 2008) for example, emphasises social resilience as important for disaster preparedness, response and post-event learning. Community competence is indicated by local understandings of risk, counselling services, absence of psychological disorders, health and wellness and quality of life. Norris et al.'s (2008) view of community resilience to disasters as a set of capacities is insightful and is shown diagrammatically in Figure 3.4. Their framework is closely allied to perspectives from public health, encompassing two dimensions of capacity: first, especially the characteristics of communities that affect their ability to identify, modify, and mobilise; and secondly, the cultivation and use of transferable knowledge, skills, systems and resources that affect community and individual changes. Thus, for Norris et al., capacities become adaptive capacities when they are 'robust, redundant or rapidly accessible and thus able to offset a new stressor, danger or surprise' (p136).

Table 3.4 Key differences between resilience and vulnerability approaches

Resilience	Vulnerability
Systems focused	Individual actor focused
Dynamic approach to managing change	Stability oriented
Multiple stressors and interactions	Specific impacts
Uncertainty about impacts	Prediction of impacts

Source: Nelson et al. 2007

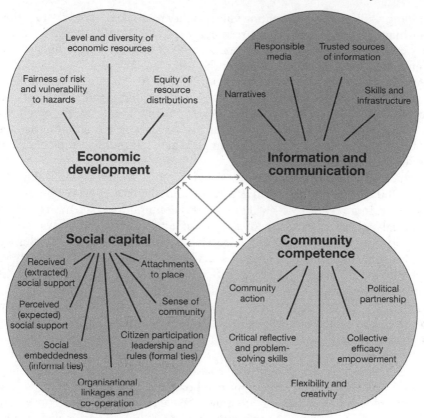

Figure 3.4 Community resilience as a networked set of capacities

Source: Brown and Westaway 2011. Originally adapted from Norris et al. 2008

It is noticeable that there is an increasing emphasis on resilience rather than vulnerability in the disasters and hazards literature. For example, Manyena (2006) notes that in the past decade work on disasters has increasingly focused on the capacity of affected communities to recover, with or without external assistance. However, he cautions that disaster resilience could be viewed as a 'new phrase describing the desired outcome of a disaster risk reduction program; but it does not deal with the unique condition itself' (Manyena 2006: 436), concluding that resilience is too vague a concept to be useful in informing the disaster risk reduction agenda. Conversely, Masten and Obradović (2008) use the findings from research in human development to argue for a resilience framework for disaster planning. Here, agency and self-efficacy are seen as important attributes that enable individuals and communities to plan, persist and adapt in the face of disasters and other events. Almedom (2009), writing from a health and social care perspective, charts a progressive paradigm shift from the disease-driven inquiries on risk and vulnerability to health-centred approaches to building resilience and preventing vulnerability to disease, social dysfunction and human and environmental resource depletion (2009). These changes can also be observed in the operations of major development and humanitarian organisations, such as the International Red Cross and Red Crescent Movement, Christian Aid and Oxfam. As Chapter 2 shows these organisations are using resilience as a key way to bring together disaster and humanitarian work with longer-term development operations. The emphasis on community resilience is also a focus of disaster risk reduction initiatives related to climate change, for example the Community and Regional Resilience Institute: ResilientUS.[4]

Much of this work emphasises social capital (Adger 2003; Pelling & High 2005) in assuming positive relationships between social capital and resilience. In a review by Rolfe et al. (2006) community resilience is related to social cohesion, in terms of the social and support networks, social participation and community engagement; social cohesion is thus a combination of social support and social capital. Chaskin (2008) analyses community resilience in three different forms – as regrouping, as redevelopment and as resistance, and relates this to community capacity manifest as the interaction of human capital, organisational resources and social capital, which can be leveraged to improve or maintain the well-being of given community.

In a recent analysis Berkes and Ross (2013) develop an integrated conceptualisation of community resilience that brings together understandings from social ecological systems, and psychology of development and mental health fields, two strands of literature they claim have been 'converging towards an appreciation of community resilience' (p6). Their model, shown in Figure 3.5, posits agency and capacity for self-organisation as the means by which different individual and collective factors, such as community infrastructure and governance (identified from across these fields) are brought together and galvanised to form community resilience. Community resilience is a function of these strengths and characteristics that build agency and self-organisation. This model is important for two reasons. First, it implies that resilience is a dynamic process, depending on the linkages or interactions between a set of different cross-scale factors or characteristics (similar in respects to Norris et al.'s [2008] linked capacities). Second, it suggests that agency and self-organisation are critical to determine how these factors might support or generate resilience. I pick up on these issues in later discussions and develop a re-visioning of resilience which puts agency – human and non-human – at its core.

A number of other recent studies, for example Buikstra et al. (2010), have developed assessments of community resilience. There is also a large literature applying a resilience lens to conceptualising and assessing adaptive capacity and social resilience (for example Maclean et al. 2014) which is discussed further in Chapter 5. Some writers challenge the ability of the community focus to adequately address multi-scalar notions of resilience

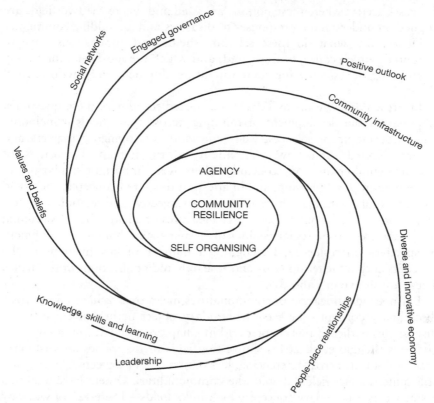

Figure 3.5 An integrated vision of community resilience

Source: Berkes and Ross 2013

(Davidson 2013; Quilley 2012), and the omission of structural factors and political dynamics from analysis of community resilience is apparent. Amundsen (2012) reminds us of some of the dangers of perceived resilience. Her analysis of one village in northern Norway warns of the 'illusion' of resilience that may lead to complacency and actually undermine or discourage adaptation to multiple change factors. In my final chapter I discuss how these differing views are reflected in experiences of Hurricane Katrina in USA.

Navigating diverse fields and applications

This chapter presents an overview of two scientific fields which inform my conceptualisation of resilience for international development and environmental change: social ecological systems, which has its roots in ecology, broadened to encompass a wider set of disciplinary perspectives from social and biosciences; and human development, which brings together psychology and child development, anthropology and sociology. It also discusses arenas where these notions are used and where the two fields are applied; in understanding responses to disasters and in building community resilience. But what do these scientific fields and applications have in common? Which concepts are shared, and which are most relevant for my re-visioning of resilience for international development and environmental change?

In Brown and Westaway (2011) we started to address these questions. We charted the development of thinking across these fields – including human development sciences, well-being and development, disasters and natural hazards and how they might inform human dimensions of environmental change. We concluded there were important similarities in the evolution of scholarship, and in addition to shared concepts, each field had undergone paradigm shifts to integrate subjective and relational aspects with more conventional and objective measures of change. In re-examining the literatures from social ecological systems and human development perspectives on resilience, I recognise a number of key overlaps in the concepts and in the tensions around research and applications of resilience. These are shown in Table 3.5.

But does identification of commonalities mean that resilience is a cross-disciplinary concept – or even a boundary object or bridging concept? I discuss this earlier in this chapter and in Chapters 1 and 2. Our analysis of citations (Baggio et al. 2015) shows the fields or disciplines are distinct – there is not much cross-referencing, less than might be expected. So they are still quite distinct fields, despite the commonalities. Others have analysed resilience as a boundary concept which links fields – I referred to work by Brand and Jax (2007) in Chapter 2. Brand and Jax explore resilience as a descriptive concept and a boundary object, emphasising the ambiguous nature of the concept. They examine how resilience is used as a descriptive concept and as a normative concept across disciplines or fields of ecological and social sciences. They highlight how the distinctions between the descriptive use of resilience – originating from ecology – become blurred

Table 3.5 Resilience concepts shared across fields

Commonalities	Tensions
Dynamics and complexity and multi-scalar nature of resilience	Distinctions between positive adaptation and transformations/bouncing back and bouncing forward
Interacting and synergistic stressors	
Transitions and thresholds	Social constructions of resilience
Windows of opportunity	Power and agency
Adaptive and other capacities	Objective vs subjective measures
	Individual traits and capacities vs collective actions and capacities
	Descriptive vs normative approaches

Source: Author's own

and often intertwined with more normative and extended uses, with individual analyses or papers often mixing multiple meanings. Thus they contend, the meaning of resilience becomes diluted and increasingly unclear in moving from a narrow ecological descriptive use to a broader normative definition, where resilience becomes a boundary object, 'floating between descriptive and normative meanings' (p32). Thus the term resilience is used ambiguously *for fundamentally different intentions* according to Brand and Jax. They propose that the increased vagueness and malleability of resilience is in fact highly valuable to foster communications between disciplines and between science policy and practice. However, they argue for what they term 'a division of labour in a scientific sense' (p33) between a descriptive resilience, a clear, well defined and measureable definition in ecological science; and social ecological resilience as a boundary object used in a transdisciplinary approach and to foster interdisciplinary work. So they recognise two quite different aspects of resilience. Other writers have highlighted this tension and confusion between resilience as a descriptive concept and resilience as a normative concept. This can be recognised in the fields and applications reviewed here – where often resilience starts as a descriptive concept, but becomes a normative concept in its applications. I do not resolve this tension, partly because I am interested in plural values and plural meanings and in how different people understand and construct meanings and experiences of resilience. My chief approach is to use resilience as an analytical lens through which to understand differentiated capacities to respond to change. I expand on this in Chapters 5 and 6. The approach I develop in this book builds on social ecological systems resilience but is informed by social science criticisms of resilience, and by the insights from the rich field of human development and the applications in disaster risk reduction and community resilience.

This overview of resilience science in these two fields reveals three important features of scholarship on resilience. First is the on-going popularity and use of the term resilience – it has power and traction in science as well as in policy and public debate. The science of resilience is diverse and dynamic, and it reflects or relates in part to the framings discussed in Chapter 1 and the discourses in Chapter 2 – but it is distinct from these. Second, resilience is grounded in different fields of scientific inquiry. I have documented the evolution of concepts and theory, the experimental and empirical testing of different theoretical perspectives on resilience and the range of methods and analysis across these fields showing its theoretical, conceptual and methodological richness (see Downes et al. 2013). Third, is the convergence around the need for greater understanding of social dynamics of resilience, and use of narratives and constructivist approaches to understand the relationships between structure and agency, and how different factors converge and will produce different outcomes for different people in different contexts. The importance of constructivist approaches, including using narratives to study peoples' accounts, experiences and stories to understand how they construct meanings of resilience is highlighted. This is further developed in the next chapter which explores plural perspectives and lived experiences of resilience.

Notes

1   This section follows the explanation provided by Resilience Alliance: www. resalliance.org/index.php/adaptive_cycle.
2   www.resalliance.org/index.php/glossary.
3   Resilience Research Centre at Dalhousie University in Halifax, Canada. See http://resilienceresearch.org.
4   www.resilientus.org.

References

Adger WN (2000) Social and ecological resilience: Are they related? *Progress in Human Geography*, 24(3), 347–64.

Adger WN (2003) Social capital, collective action, and adaptation to climate change. *Economic Geography*, 79(4), 387–404.

Almedom A (2009) Resilience research and policy/practice discourse in health, social, behavioral, and environmental sciences over the last ten years. *African Health Sciences*, 8(Suppl), S5–S13.

Amundsen H (2012) Illusions of resilience? An analysis of community responses to change in northern Norway. *Ecology and Society* 17(4), 46.

Baggio J, Brown K and Hellebrandt D (2015) Boundary object or bridging concept? A citation network analysis of resilience. *Ecology and Society*, 20(2), 2.

Bahadur AV, Ibrahim M and Tanner T (2010) *The resilience renaissance? Unpacking of resilience for tackling climate change and disasters, SCR Discussion Paper 1.* Institute of Development Studies. Available from: http://bit.ly/1cIBONB (accessed 23 October 2013).

Berkes F and Ross H (2013) Community resilience: Toward an integrated approach. *Society and Natural Resources*, 26(1), 5–20.

Berkes F, Folke C and Colding J (1998) *Linking social and ecological systems: Management practices and social mechanisms for building resilience.* New York: Cambridge University Press.

Boyden J and Cooper E (2007) *Questioning the power of resilience. Are children up to the task of disrupting the transmission of poverty?* Chronic Poverty Research Centre Working Papers.

Brand F and Jax K (2007) Focusing the meaning(s) of resilience: Resilience as a descriptive concept and a boundary object. *Ecology and Society*, 12(1), 23.

Brown K (2014) Global environmental change I: A social turn for resilience? *Progress in Human Geography*, 38(1), 107–17.

Brown K and Westaway E (2011) Agency, capacity, and resilience to environmental change: Lessons from human development, wellbeing, and disasters. *Annual Review of Environment and Resources*, 36(1), 321–42.

Buikstra E, Ross H and King C (2010) The components of resilience: Perceptions of an Australian rural community. *Journal of Community Psychology*, 38(8), 975–91.

Camfield L and McGregor A (2005) Resilience and well-being in developing countries. In: Ungar M (ed.), *Handbook for working with children and youth:*

Pathways to resilience across cultures and contexts, Thousand Oaks, CA/London/ New Delhi: SAGE, 189–209.

Carpenter S, Walker B, Anderies J and Abel, N (2001) From metaphor to measurement: Resilience of what to what? *Ecosystems*, 4(8), 765–81.

Chaskin RJ (2008) Resilience, community, and resilient communities: Conditioning contexts and collective action. *Child Care in Practice*, 14(1), 65–74.

Cutter S, Barnes L, Berry M and Burton C (2008) A place-based model for understanding community resilience to natural disasters. *Global Environmental Change*, 18(4), 598–606.

Davidson DJ (2013) We still have a long way to go, and a short time to get there: A response to Fikret Berkes and Helen Ross. *Society and Natural Resources*, 26(1), 21–4.

Downes B, Miller F and Barnett J (2013) How do we know about resilience? An analysis of empirical research on resilience, and implications for interdisciplinary praxis. *Environmental Research Letters*, 8(1), 014041.

Folke C (2006) Resilience: The emergence of a perspective for social–ecological systems analyses. *Global Environmental Change*, 16(3), 253–67.

Gallopín GC (2006) Linkages between vulnerability, resilience, and adaptive capacity. *Global Environmental Change*, 16(3), 293–303.

Gunderson LH and Holling CS (eds) (2002) *Panarchy: Understanding transformations in human and natural systems*. Washington, Covelo, London: Island Press.

Hall PA and Lamont M (2013) *Social Resilience in the Neoliberal Era*. Cambridge: Cambridge University Press.

Holling CS (1973) Resilience and stability of ecological systems. *Annual Review of Ecology and Systematics*, 4(1), 1–23.

Holling CS (1978) Adaptive environmental assessment and management. In: Holling CS (ed.), *Adaptive environmental assessment and management*, London: Wiley, p. 377.

Janssen M, Schoon M, Ke W and Börner K (2006) Scholarly networks on resilience, vulnerability and adaptation within the human dimensions of global environmental change. *Global Environmental Change*, 16(3), 240–52.

Kaplan H (1999) Toward an understanding of resilience. In: Glantz M and Johnson J (eds), *Resilience and development: Positive life adaptations*, New York: Kluwer Academic, 17–83.

Leichenko R and O'Brien K (2008) *Environmental change and globalisation: Double exposures*. Oxford: Oxford University Press.

Ludwig D, Jones DD and Holling CS (1978) Qualitative analysis of insect outbreak systems: The spruce budworm and forest. *The Journal of Animal Ecology*, 47(1), 315–32.

Luthar SS (2006) Resilience in development: A synthesis of research across five decades. In: Cicchetti D and Cohen DJ. (eds), *Developmental psychopathology: Risk, disorder and adaptation*, New York: Wiley, 740–95.

Luthar SS and Cicchetti D (2000) The construct of resilience: Implications for interventions and social policies. *Development and Psychopathology*, 12(04), 857–85.

Maclean K, Cuthill M and Ross H (2014) Six attributes of social resilience. *Journal of Environmental Planning and Management*, 57(1), 144–56.

Manyena SB (2006) The concept of resilience revisited. *Disasters*, 30(4), 433–50.

Martin-Breen P and Anderies J (2011) *Resilience: A literature review: The Rockefeller Foundation*. Brighton: IDS: Bellagio Initiative. Available from: http://bit. ly/1Hhz2sy (accessed 27 February 2013).

Masten A (2001) Ordinary magic: Resilience processes in development. *American Psychologist*, 56(3), 227.

Masten A (2007) Resilience in developing systems: Progress and promise as the fourth wave rises. *Development and psychopathology*, 19(3), 921–30.

Masten A, Best KM and Garmezy N (1990) Resilience and development: Contributions from the study of children who overcome adversity. *Development and Psychopathology*, 2(4), 425–44.

Masten A, Cutuli JE, Herbers JE and Reed M (2013). Resilience processes in development: four waves of research on positive adaptation in the context of adversity. In SJ Lopez and CR Snyder (eds.), *Handbook of Resilience in Children* (pp. 15–38). Boston, MA: Springer US. doi:10.1007/978-1-4614-3661-4.

Masten A and Obradović J (2008) Disaster preparation and recovery: Lessons from research on resilience in human development. *Ecology and Society*, 13(1), 9.

Masten A and Reed M (2002) Resilience in development. In: Lopez SJ and Snyder CR (eds), *The handbook of positive psychology*, New York: Oxford University Press, 77–88.

Nelson D, Adger WN and Brown K (2007) Adaptation to environmental change: Contributions of a resilience framework. *Annual Review of Environment and Resources*, 32(1), 395–419.

Norris F, Betty SP, Pfefferbaum B, Wyche K and Pfefferbaum R (2008) Community resilience as a metaphor, theory, set of capacities, and strategy for disaster readiness. *American Journal of Community Psychology*, 41(1–2), 127–50.

O'Dougherty Wright M and Masten A (2005) Resilience processes in development. In: *Handbook of resilience in children*, New York: Springer, 17–37.

Pelling M and High C (2005) Understanding adaptation: What can social capital offer assessments of adaptive capacity? *Global Environmental Change*, 15(4), 308–19.

Quilley S (2012) *Resilience through relocalisation: Ecocultures of transition? Ecocultures Working Paper: 2012–1*. UK: University of Essex. Available from: http://bit.ly/1A2NLs2 (accessed 14 May 2015).

Rolfe RE, Langille D, Maher R, Stewart DM, Pilkey D, Smith M, Vanderplaat M and Stiles D (2006) *Social cohesion and community resilience: A multi-disciplinary review of literature for rural health*. Available from: http://bit.ly/1PINVX6.

Rutter M (1979) Protective factors in children's responses to stress and disadvantage. In: Kent MW and Rolf JE (eds), *Primary prevention of psychopathology* (*Social competence in children, vol. 3, pp. 49–74*), Hanover: NH: University Press of New England.

Rutter M (2000) Resilience reconsidered: Conceptual considerations, empirical findings, and policy implications. In: Shonkoff JP and Meisels SJ (eds), *Handbook of early childhood intervention*, New York: Cambridge University Press, 651–82.

Schoon I (2006) *Risk and resilience: Adaptations in changing times*. Cambridge: Cambridge University Press.

Ungar M (2004) A constructionist discourse on resilience multiple contexts, multiple realities among at-risk children and youth. *Youth & Society*, 35(3), 341–65.

Ungar M. (ed.) (2005a) *Handbook for working with children and youth: Pathways to resilience across cultures and contexts*. London: SAGE Publications.

Ungar M (2005b) Resilience among children in child welfare, corrections, mental health and educational settings: Recommendations for service. *Child and Youth Care Forum*, 34(6), 445–64.

Ungar M, Brown M, Liebenberg L, Othman R, Kwong W, Armstrong M and Gilgun J (2007) Unique pathways to resilience across cultures. *Adolescence*, 42(166), 287–310.

Walker B and Salt D (2006) *Resilience thinking: Sustaining ecosystems and people in a changing World*. Washington, Covelo, London: Island Press.

Walker B, Holling CS, Carpenter SR and Kinzig, A (2004) Resilience, adaptability and transformability in social–ecological systems. *Ecology and Society*, 9(2), 5.

Walker BH, Ludwig D, Holling CS and Peterman RM (1981) Stability of semi-arid savannah grazing systems. *Journal of Ecology*, 69(2), 473–98.

Xu L and Marinova D (2013) Resilience thinking: A bibliometric analysis of socio-ecological research. *Scientometrics*, 96(3), 911–27.

4   Exploring experiential resilience

Resilience has multiple and often contested meanings. So far this book has looked at how different scientific traditions and fields understand resilience, and how resilience ideas are being used in policy arenas related to international development and environmental change, by exploring the framings, policy discourses and different scientific perspectives on resilience. This chapter presents a quite different view of resilience. In this chapter I ask the question, *how is resilience lived and experienced?* This chapter brings resilience 'down to earth' and examines some of the key issues which people highlight when explaining their vulnerability, and their strategies and capabilities to deal with change. It relates these narratives of resilience and the stories that people construct about resilience. In doing this, it exposes the social and political dynamics of experiential resilience.

The chapter employs a series of vignettes[1] to study how people have responded to different shocks and changes in very diverse contexts, including East and Central Africa, and the UK. I use these to explore five dimensions of resilience highlighted by political ecology approaches and hitherto overlooked by much resilience research, which were introduced in Chapter 1. The vignettes highlight the importance of power asymmetries that influence peoples' capacity to respond to change. This is especially evident in how resilience is often articulated as closely aligned with resistance or empowerment. Second, they draw attention to cross-scale interactions and the impacts of interventions and top-down policies on resilience, especially of the poor. Third, they highlight how resilience and vulnerability play out for different individuals and households, and illustrate the socially differentiated nature of resilience, mediated by a range of factors and intersecting stressors, such as weather extremes, changes in policy and ill-health. Fourth, the contested nature of values and knowledges underpinning resilience and how this contestation affects the dynamics of change, and marginalised and indigenous peoples' capacities to respond is shown. Fifth, the vignettes reveal the importance of 'situated' resilience and how a sense of place plays an important role in how different social actors are affected by and effect change. Finally I use these 'lived experiences' and narratives of resilience to synthesise some core themes for resilience insights for international development and environmental change and to signal the importance of 'everyday forms of resilience'. As these issues are not well articulated in literature on social ecological systems resilience, and only partially addressed in other resilience literatures, the first section offers some background and an overview of social vulnerability that informs these vignettes.

Understanding vulnerability and resilience

In Chapter 1 I outlined a broadly political ecology approach to resilience, emphasising the analysis of discourse, narratives, competing knowledge claims, power and cross–scale dynamics. Using this analytical lens extends conventional approaches and insights, and develops a more socially and politically nuanced understanding of human capacities and opportunities in the face of change. This expands and compliments current approaches in the social ecological systems resilience field to make it speak more usefully to international development and environmental change. The five themes explored in this chapter, although under-emphasised to date in resilience literature, are well represented in work by geographers, political scientists, development studies scholars and others. Analysing these dimensions leads to a more socially and politically nuanced understanding of resilience, and a more critical social science of resilience to inform development and change as outlined in later chapters. A political ecology approach to resilience makes explicit the questions, who is resilient, and what are they resilient to? Furthermore, taking this approach recognises that there are important trade-offs in terms of different peoples' welfare or well-being associated with resilience.

The work of political ecologists, political and social geographers, and scholars concerned with international development and social vulnerability has highlighted the socially differentiated nature of environmental change impacts (Leichenko & O'Brien 2008). Increasingly, an environmental justice lens is applied to understanding how the different costs and benefits, and access and voice are played out in dynamic and reflexive responses to environmental and other changes. Climate change adaptation in particular has become the subject of analysis from a justice perspective (Adger et al. 2006). Mearns and Norton (2009) apply a social justice lens, highlighting how climate change acts as a multiplier of existing vulnerabilities; both causes and consequences of climate change are deeply intertwined with global patterns of inequality. Their analysis emphasises the necessity for climate change adaptation policies to consider and incorporate gender issues and the voices of marginalised groups including the poorest and indigenous people.

Much of the literature on social vulnerability discusses the socially differentiated nature of peoples' capacity to cope with change and particularly with shocks such as extreme weather events or disasters. For example, Brouwer and Nhassengo's (2006) analysis of devastating floods of 2000 in Limpopo Province in Mozambique demonstrates how vulnerability cannot be simply equated with poverty; it is more fluid, complex and specific to time, space and social context. In an extreme event such as these floods, many factors determine who is affected and how. In this case, it was comparatively richer households – those with cattle – who lost their houses and stock, and also it was the richest who were least likely to be helped by traditional support mechanisms in the community. As the authors conclude, 'Although these mechanisms have made communities ... highly resilient to stresses, such as drought and floods, they are apparently not strong enough to deal with a disaster on a scale like that of the floods of 2000' (Brouwer & Nhassengo, 2006: 251), emphasising the significance of the nature of the event itself.

Social vulnerability analysis gives insights then into how and why different social groups, or even different individuals, are affected differently by change and have different capacities to respond. Ribot (2009) provides an insightful review of approaches to vulnerability analysis, distinguishing between two polarised archetypes. A risk-hazard approach tends to evaluate multiple outcomes or impacts of a single event. A social constructivist approach characterises the multiple causes of single outcomes. The first approach essentially uses a dose-response relation between an exogenous hazard to a system and its impacts, locating risk within the hazard itself. The second approach in contrast, seeks to understand what causes vulnerability; although a climate related event may be an external phenomenon, risk is located in society itself – for example, Blaikie et al. (1994) see access to resources as a critical determinant shaping peoples' vulnerability. As Ribot explains (2009) the development of integrated frameworks that view vulnerability as depending on both biophysical and human factors is necessary within the environmental change and climate change literature. Ribot views the most informative of these – essentially extensions of social or constructivist approaches to vulnerability – as tracing the causes of vulnerability from specific instances of risk in order to explain why a given individual, household, group, nation or region is at risk from a particular set of events or damages. This presents an integrative, causality-based understanding, which allows for a multi-scale and multifactor analysis of vulnerability.

The terms vulnerability, risk and resilience are often used together, and are not very clearly distinguished in literature on hazards and disasters. Indeed, some authors suggest that resilience is an antonym of vulnerability. How then is this vulnerability approach distinct from a resilience one, and how can it inform and enrich resilience-based approaches? Resilience and vulnerability are not straightforward opposites as Table 3.4 shows, and there are complementarities and convergences in their applications; they are conceptually closely allied (see Miller et al. 2010). Nelson et al. (2007) suggest that vulnerability, like resilience, is another property or characteristic of a social ecological system. Berkes (2007) more specifically suggests that resilience thinking provides a more dynamic and integrated systems thinking approach to understanding hazards. Almedom and Tumwine (2008) view a shift from an emphasis on vulnerability to one of resilience in the disasters field as a move away from deficit models to ideas about assets and strengths, representing a 'paradigm shift from vulnerability to strength'. But a more socially-informed and nuanced resilience approach benefits immensely from studies of vulnerability, and especially so from analysis which highlights the socially differentiated nature of vulnerability and its multiple, often interlocking and structural causes. In this respect the literature on social vulnerability is particularly significant for a political ecology approach to resilience.

Empirically-informed analyses of vulnerability and adaptation can very helpfully inform the political dynamics of resilience. For example, Eriksen and Lind's (2009) analysis of drought and conflict in Kenya highlights adaptation to change as an 'intrinsically political process' (p817). Pastoralists and dryland farmers form strategic alliances with more powerful political elites to influence collective decision-making as well as to gain access to resources. Whilst adjustments to drought are related to on-going conflicts over resources, particularly access to land, systematic regional biases in development policies and allocation of resources further disadvantage pastoralists. Although there are winners and losers in adaptation, whether or not adjustments to drought exacerbate existing inequalities depends on the power relations that influence the shape and outcome of negotiations. O'Brien cand Leichenko (2003) highlight that even though the terms 'winners' and 'losers' are used frequently in discussions of global change, the insights gained from empirical analysis and from understanding the processes which create winners and losers are more rarely analysed. Therefore, terms like winners and losers are applied, without fully understanding the processes that create them. In taking a systems approach, a resilience lens might provide insights into these processes, although to date few studies consider these dynamics.

In discussing the responses to earthquakes in Christchurch, New Zealand, Hayward (2013) suggests that 'rethinking resilience' should place politics much more centrally in analysing responses to both environmental and other stressors. Despite its roots in complex adaptive systems theory, resilience science rarely sees vulnerability – or a lack of resilience – as being produced by social processes. The resilience lens – often, but not always – focuses attention on shocks as being external to the system. Because much resilience research focuses on place-based analysis of particular social ecological systems, this often obscures the way in which economic and power relations are privileged. In turn this leads to underestimating the extent to which transformations in social ecological systems and governance actually require transformations of power and concerted political struggle.

Power is also closely allied and linked to knowledge and to access to resources. There are some interesting strands in the literature on environmental change, resilience and adaptation concerning different forms of knowledge and the interplay between indigenous and scientific knowledge. Knowledge contestations are highlighted in political ecology studies of natural resources, environment and policy. This further underscores the importance of viewing knowledge as dynamic, and to recognise the interplay between different forms of knowledge.

To move beyond a snapshot view and understand the processes that create and recreate vulnerability and resilience over time requires longitudinal studies, which are relatively rare in resilience studies. This chapter introduces three key longitudinal studies. Another longitudinal study is provided by Smucker and Wisner (2008) analysing change over more than 30 years in Tharaka in Kenya. This study shows that major macro-level transformations – including population growth, privatisation of land ownership, decentralisation, environmental change, market shifts and increasing conflicts over resources have undermined people's capacity to cope with drought. Many of the strategies employed recently have been incompatible with long-term secure livelihoods. Although the authors claim that people in Tharaka are resilient, their analysis contradicts this, highlighting a narrowing of options, coping strategies and increased dependence on food aid. As in the cases discussed here from Tanzania and Mozambique (and also observed by Eriksen and Silva [2009]) many of these macro-level changes associated with conventional development (privatisation of land tenure, increased market integration) have altered people's capacity to respond to drought and other contingencies. As further explored in later chapters, external interventions and development assistance may have maladaptive consequences and may increase vulnerability to climate change and other changes.

The empirical investigations discussed in the vignettes here each emphasise the concerns that people raise when they talk about their resilience. The studies reveal what different people regard as important for resilience, and what they highlight as critical for their own capacity to respond positively to change. Analysing narratives of change in terms of vulnerability, adaptation and resilience thus provides rich storylines to understand different perspectives and ultimately to support action and policy. They give strength to the value of constructivist approaches to understanding resilience as highlighted in the recent literature on resilience in human development. These narratives can be seen as a means of explaining change processes and responses, and to gain insights into diverse meanings and mental models. They ultimately constitute a translation of complexity rooted in the lived experience of different people, in different places. Table 4.1 summarises the five vignettes and the resilience dimensions they highlight.

Table 4.1 Five vignettes of experiential resilience

Case	Resilience dimensions
Economic crisis and change in southern Cameroon Brown & Lapuyade 2001	Intra-household winners and losers; gendered capacities and responses; inter-acting multiple stressors; social and environmental dimensions of change.
Impacts of HIV/AIDS on rural households in Uganda Seeley et al. 2009	Inter- and intra-household and inter-generational differences in coping capacity; household resilience and the 'costs' of recovery; 'Bouncing back' relationship obscures social costs, winners and losers.
Climate hazards and vulnerability in coastal Mozambique and Tanzania Bunce et al. 2010a	Inter-household difference, poverty and lack of access to key resources and assets, and powerlessness; role of conventional development in undermining resilience of the poor.
Living with change in Orkney Few et al. 2007	Narratives of change, resistance and resilience; self-determination and autonomy; cultural values and rootedness, sense of place.
Indigenous resilience Rotarangi & Russell 2009	Marginalisation, cultural and political dynamics, knowledge differences and contestations, place attachment.

Source: Author's own

'Maybe this is the end of the world': women and men, winners and losers

This quote from a woman farmer in southern Cameroon expressed the catastrophic effects of economic crisis on households, articulated in field research and reported in Brown and Lapuyade (2001). The economic crisis in Cameroon in the mid-1990s hit many sectors of the economy and had a series of inter-related impacts on urban and rural dwellers, on agriculture, forests, and livelihoods, opportunities and aspirations. Our research in southern Cameroon in 1999 and 2000 was able to build on earlier research by Jane Guyer, who worked in southern Cameroon from the 1960s. Our research aimed to gain insights into the socially differentiated impacts of a boom and bust economy, and in turn, how this affected forest use. In the 1970s Guyer (1984) had shown that the role of women had shifted from subsistence maintenance in a colonial and cash crop economy, to the provision of food for the urban market during the post-colonial period. Her work during this period showed that the gap in status between men and women narrowed; principally due to a decline in male status, and that Beti women (from southern Cameroon) in general thought that their situation had improved compared with the pre-colonial and colonial times. However, between the mid-1980s and mid-1990s, the combined impacts of economic crisis and Cameroon's structural adjustment programme (SAP) resulted in a series of changes that had impacts across the country and upon all sectors of the economy. A steady period of economic growth (based on oil exports, expansion of cash crops and a timber boom) between 1960 and 1985 was brought to a sudden halt between 1985 and 1986. An economic recession, sharp decrease in the price of export crops, devaluation of the CFA franc and the general disengagement of the state had severe repercussions in both rural and urban society (Brown & Lapuyade 2001). Women and men interviewed in 1999 gave contrasting perspectives on how these changes had affected their lives. Life was perceived to have been especially difficult for women because of their greater need for cash to support households and children, and the poor terms of trade they received for their produce. Whereas men still regarded their standard of living as steadily improving, women's livelihoods, sources of income and work burdens were especially sensitive to increases in prices of food and fuel; consumables cost more, whilst prices for products they sold had fallen.

These economic and social changes presented both opportunities and constraints for villagers in Komassi, the site of our in-depth study in southern Cameroon. They resulted in shifts in responsibilities and impacted on relative incomes. What was particularly important in exacerbating the impacts of these changes on women was their differentiated capacity to adapt to changes and shocks. Men were able to diversify their sources of livelihood in response to external and internal changes, especially by shifting from cash crops to growing food crops. On the other hand, women were less able to call on additional income sources and had increasingly commercialised their livelihood activities to overcome cash shortages. In addition to more food cropping, men also planted coffee (to replace cocoa), and had access to new sources of income through crafts or trade, and pensions. Increasingly it was women who felt the impacts of these changes – especially to their traditional sources of income generation from food crops. Coupled with deteriorating terms of trade for their food crops and higher fuel costs, women undertook more food processing. Meanwhile, women linked these developments to higher work burdens and lower returns, and to their increasing cash needs – as SAP policies hit rural services such as health services and schools. Furthermore, it appeared that as women's ability to generate income was reduced, their power to re-negotiate their roles and responsibilities also deteriorated (Brown & Lapuyade 2001).

Of course these changes also had important impacts on villagers who were able to access and use natural resources, especially forest resources. As one elderly woman told us, 'the crisis has fallen on the forest' – meaning that people exploited and sold a far greater amount and variety of non-timber forest products (NTFPs) and that products were becoming increasingly difficult to find. Poaching was also perceived to have increased with the economic crisis. A number of studies have linked increased rates of deforestation to the economic crisis and the migration of people back to rural areas (Sunderlin et al. 2000). Forest clearance was extended into areas further from the village (Brown & Lapuyade 2001). But women and men also noted that new rules were emerging to control access to forest resources. We concluded that the increased trade in NTFPs was seen, at least partially, as negative for women and children. Women's traditional rights to forest resources were being eroded through competition with migrants and with men, thus further constraining their capacity to adapt to changes and diversify their livelihoods.

So, men and women had different capacities and capabilities to deal with the changes precipitated by economic crisis. Whilst men could exploit these changes and find new economic niches and opportunities to earn cash, structural constraints, unequal access to resources and decision-making, and greater burdens of labour and caring meant women coped in different ways and lost out – in effect they were less resilient. This case shows how the interlocking and inter-related effects of economic, social and environmental change had differentiated impacts even on people (men, women and children) within the same households.

'This is where we buried our sons': social impacts of HIV/AIDS in Uganda

This quote by an elderly widow in southwest Uganda is used in the title of an article by Janet Seeley and colleagues (2009: 115) detailing findings from an ethnographic study of the impact of HIV and AIDS on households in Uganda.

Seeley and colleagues at the joint Medical Research Council (MRC) and Uganda Virus Research Institute (UVRI) Uganda Research Unit on AIDS have been examining the impact of HIV and AIDS on rural societies in Uganda and Tanzania since the 1980s. Their analysis provides unique insights into the trade-offs between different social actors and across time and space inherent in resilience. The region and communities they studied have been described as an epicentre of the global AIDS pandemic, with high infection rates and a highly susceptible population. AIDS was first identified there in 1980s, and the MRC/UVRI research programme has followed a large cohort of households over a 30-year period. They have also undertaken in-depth ethnographic research with 24 individuals and been able to map very diligently how individuals have coped and fared, and the personal narratives and realities of coping with the ravages of ill health and bereavement in families, alongside other stressors and changes.

This longitudinal study was able to quantify the long-term impacts of households experiencing an AIDS-related death of an adult member (Seeley et al. 2010). On average, household consumption dropped seven per cent within in the first five years after an adult death. However, this impact did not persist and after six years, the consumption by families that had experienced an adult death was no different from families that had no experience of such an event. Figure 4.1 shows this apparent 'bounce back' after the death of a household member in terms of land cultivated by households.

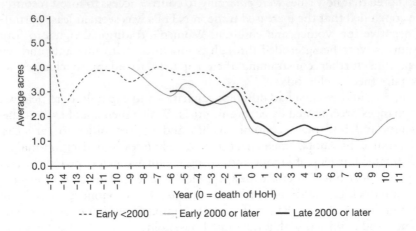

Figure 4.1 Bouncing back: impact of death of household head on cultivation

Source: Seeley, Dercon and Barnett 2010

Seeley et al. (2010: 333) remark that 'from today's perspective, the AIDS epidemic seems not to have had the profound long-term impact in this part of East Africa that was predicted 20 years ago'. The households and communities have displayed unanticipated resilience. But the ethnographic evidence they present shows the devastating impacts on individual and households. As they explain (p333): 'children are growing up without parents and grandparents, and individuals have experienced multiple and profound hardships associated with growing up an orphan in poverty and deprivation.' Clearly the life opportunities for these individuals had been severely constrained by the epidemic and their physical and emotional well-being undermined. Although Ugandan communities appear resilient, the development costs have been profound and will affect generations to come.

Although much literature focused on the impacts upon the young (as orphans) and adults (as victims of disease), Seeley and colleagues' longitudinal study reveals that the elderly – and the oldest – are particularly vulnerable. In this sense some of the impacts were displaced temporally (Seeley et al. 2009). People aged over 75 years have often had to take on the roles of a younger relative (childcare, responsibility for schooling, farming), had they lived. In this study, the elderly faced a potential crisis in having no-one to care for them as they aged and became less independent. The elderly with fragile health and who were also poor were especially at risk. HIV/AIDS is just one of the many challenges faced by households, with poverty and ill heath critically important. Ill health in particular erodes peoples' wealth and their opportunities (see also Terry 2011). Poverty exacerbates the losses experienced as a result of AIDS and socio-economic status and family size are key determinants of ability to 'bounce back'. Seeley et al.'s in-depth ethnographic analysis presented in their 2009 paper also highlights the significance of social relations. A person's life is embedded across the lifespan in social relationships with relatives, friends and neighbours, and this also influences coping capacity.

There are important lessons for resilience from these findings. A major one concerns the role of scale, and about how we understand and assess resilience – over what spatial or social scale, and over what time period. The aggregated analyses show a clear 'bounce back'; the mean time it took households to recover from an AIDS-related death – measured by the size of the household and the amount of land cultivated – was five or six years. But these aggregations hide the true cost and the suffering at personal and household levels, which are revealed in the ethnographic study. The poor, the elderly and the young in particular bore the costs especially in missed opportunities and immediate well-being. The socially differentiated impacts and the social costs of resilience are somewhat hidden in the aggregated analyses but are critically important for development and well-being within communities.

'We don't own the future': coastal vulnerability in East Africa

This comment from a fisher in Mozambique captures the powerlessness and vulnerability expressed in interviews we carried out in Tanzania and Mozambique to understand how poor people perceive their resilience to climate change (Bunce et al. 2010a). This research suggests that many East African coastal inhabitants consider themselves vulnerable; they face multiple stresses which bring mainly negative changes to their livelihoods and security, and generally feel powerless and helpless in the face of change. As this fisher summed up these feelings: 'We don't know what will happen. Up to now no-one has come to help us. We don't own the future. There is no communication about what is happening' (Bunce et al. 2010a: Table 4, p491). This remark also points to a failure of contemporary development policies to improve the security of the most vulnerable. Although people ranked climatic variability as being very important, these were not the most important changes which affected peoples' livelihoods, as shown in Table 4.2.

Compiling simple mental models of vulnerability and resilience reveals how people make linkages between a number of different stressors and how they see a range of different changes affecting them. Very often it is the interactions between stressors which combine to cause the greatest impacts (Bunce et al. 2010b; Brown & Ekoko 2001). For example, the effects of drought, rising food prices and upstream river basin management in Mozambique were seen as combining to impoverish smallholder farmers and fishers. People resorted to skipping meals as a result; risking malnutrition and making them more vulnerable to a range of diseases. As with the case from southwestern Uganda, the interactions of poverty, ill health and other stresses have severe impacts on peoples' welfare and ability to cope with change.

Table 4.2 Ranking of events and changes in coastal Mozambique and Tanzania

Mozambique (n = 13)	Mentions	Tanzania (n = 15)	Mentions
Food prices (rising)	11	Less fish	12
Less rain (infrequent/erratic)	8	Less rain	10
River floods (frequency/severity)	8	Rising illness	9
		Food prices (rising)	9
Rising illness	7	Low crop prices	6
Winds stronger	6	Less crops	5
Temperature rising/drought	5	Fewer jobs	5
Less fish catch	4	Population rise/density	5
Soil salinity (river)	3	Soil depletion	5
Sea flooding (tide heights/surges)	3	Lower fish prices (sell)	2
		Less credit access	2
Population rise/density	2		
War	2		

Source: Bunce et al. 2010a: 424

But what our work in East Africa identified especially was that conventional conservation and development policies were additional stresses and actually undermined peoples' resilience. In Macaneta at the mouth of the Incomati River in Mozambique, villagers identified upstream river basin management as a root cause of loss of livelihood and income (see Figure 4.2). Upstream dams exacerbate water stress, as one donor representative we interviewed explained, 'When it is dry South Africa keeps too much and when it rains they let it slosh into Mozambique' (Bunce et al. 2010b: 492). The demands for water upstream by commercial agriculture and the need to maintain supplies to urban areas, including Maputo, disadvantage the rural smallholders at the mouth of the estuary. Their vulnerability and flood risk are increased. In this case vulnerability and resilience have important geographical and transboundary dimensions.

In Mnazi Bay in Tanzania, the establishment of a marine protected area (MPA) in 2000 fuelled peoples' sense of vulnerability and of disempowerment. As shown in Figure 4.3, the MPA was seen as contributing to loss of livelihood by reducing the area where people could fish and forcing people to fish further from shore. This was exacerbated by the land use restrictions associated with the protected area designation. The objectives of the MPA included 'rational development of under-utilised natural resources, and the management of marine and coastal areas so as to promote sustainability of existing resource use, and the recovery of areas and resources that have been over-exploited or otherwise damaged' (Bunce et al. 2010b: 490). However, certain policies specifically designed to increase sustainability were seen to undermine the resilience of poor people from certain parts of the

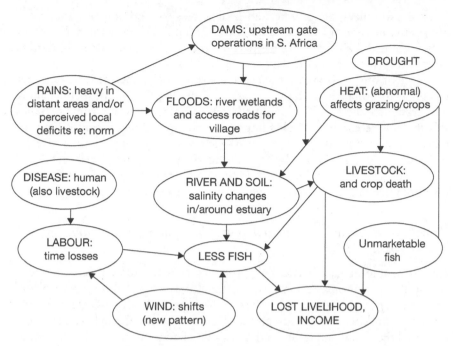

Figure 4.2 Mozambique villager mental model of impacts of water management

Source: Bunce et al. 2010b: 492

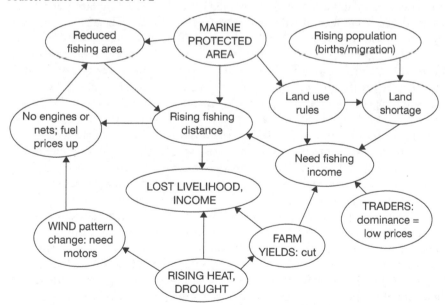

Figure 4.3 Tanzania villager mental model of impacts of marine protected area

Source: Bunce et al. 2010b: 490

coast – particularly those who had little access to land and few sources of alternative income or livelihood for whom extending fishery further offshore was not possible. These were likely to be those with fewer assets – in other words, they were the poorest.

Many of the stressors are themselves a result of deliberate interventions, designed to enhance welfare, though not necessarily of the people more vulnerable to change. We identified this as comprising a form of 'policy misfit'. This research highlights important synergistic relationships between stressors, and how they acted at different scales – many of them are far beyond the local. Second, by examining peoples' own experiences, their narratives and their mental models, it demonstrates that people felt very disempowered to act and to influence these stressors.

'Local people know best how to manage these islands': narratives of coastal change in Orkney

Just as people in East Africa link powerlessness to vulnerability, so when people talk about their resilience it is often deeply rooted in their perceptions of self-determination and self-efficacy. My research with Roger Few and Emma Tompkins and others at the Tyndall Centre (Brown et al. 2005) explored the social and political dimensions of adaptation to climate change impacts in the coastal areas of the Orkney islands of the north coast of Scotland (Few et al. 2007a; Tompkins et al. 2008).The quote above is from a participant in a workshop we convened on Orkney looking at how we can make decisions in face of climate change (Few et al. 2007b: 52), but it sums up the mood in Orkney for local-level autonomy over planning. Our research sought to answer not what *should* be done, but *how ought decisions be made* in the face of uncertainty about timing and severity and precise location of future impacts of climate change. In Orkney, the work centred around the potential for disruption of transport within Orkney and between the islands and mainland Scotland as a result of potential sea level rise and increases in storminess. A series of interviews and workshops were undertaken with a range of stakeholders, who included local councillors, fishers and communities. Participatory scenarios and deliberative polling were used to explore present and future priorities and trade-offs. By discussing the pros and cons and the trade-offs between centralised and decentralised approaches, and between anticipatory or delayed action issues of local autonomy, power and politics, and current and future vitality, sustainability and resilience were voiced. The nature of resilience of communities on Orkney was revealed through these narratives, which spoke directly to issues of self-sufficiency and resistance. Some quotes illustrating how people expressed these concepts are shown in Box 4.1.

From these workshops a strong narrative about perceived resilience of Orcadians emerged – as island folk with a strong seafaring tradition and history of resistance to both colonising forces and challenging environmental

---

Box 4.1 Resilience and self-sufficiency in Orkney

'Generally people don't see weather or movement of the sea as a problem. It's something to accommodate, accept and work around.' Local councillor, Stromness

'Orcadians should think ahead to alleviate prospective problems and need to start planning now.' Librarian, Kirkwall

'Orkney may need to become more self-sufficient in many food products to reduce dependence on the importation of stocks. The threat of disruption of transport may have a positive effect in stimulating more local production and more local trades.' Fisher, Sromness

'Orcadians are resilient and should draw on their tradition of pragmatism and adaptability to look upon climate change as an opportunity not just a threat.' Workshop participant, Stromness

Source: Adapted from Brown et al. 2005

---

conditions. A certain pride was expressed about this heritage, and the uniqueness of the island culture, physical and especially archaeological assets. People were also sensitive to the various threats to sustainability and future of the islands, most of which did not relate to climate change. The very things that made Orkney special, were also at the root of its potential decline: sustaining local communities, attracting incomers as well as providing incentives for people to remain – especially on outer islands – were current and future priorities. Remoteness was already a disincentive for incomers, and threatened social and economic vitality. Thus a series of quite conflicting discourses was uncovered, and the discussion of different scenarios for future decision-making was a good opportunity to explore these apparent contradictions and differing opinions.

This process revealed, on the one hand, a strong sense of optimism, a belief in the resilience and adaptability of Orcadians. The idea that shocks to the system may prove to be 'windows of opportunity' (see Folke 2006, discussed in Chapter 3) was strongly articulated. Some people felt that easy transport between islands and to mainland Scotland had undermined long-term resilience and self-sufficiency, that communities had become dependent on these links and on goods and services from the mainland.

On the other hand, the economic dependence on mainland Scotland was recognised and the participants were both realistic about this and also creative about new ways to generate income, particularly associated with renewable energy. People had lots of ideas about how anticipatory action could be taken without major investments, articulating an adaptive management approach very clearly. Box 4.2 summarises the trade-offs and the types of shocks as articulated by different people on the islands during the workshops.

In Orkney there was an expressed desire to have more local autonomy in decision-making, but also a recognition that if external funding were to be obtained (which was necessary) then some centralised control is inevitable. But Orcadians have a strong commitment to local democracy to represent the best interests of islanders. This case emphasises the central importance of self-determination and voice that links peoples' ideas of resilience to justice. Resilience was articulated here as self-determination, as self-sufficiency and autonomy. It was also critically linked to cultural identity and sense of place, as well as to history and heritage. I discuss these aspects much more in Chapter 7, where I present a revised view of resilience putting resistance, rootedness and resourcefulness at its core.

'To be indigenous is to be resilient': resilience at the margins

This close alignment between resilience and resistance is strongly echoed in literature on indigenous people, environmental management and social,

---

Box 4.2 Making decisions about climate change in Orkney

'Investment in planning for possible climate impacts might be better spent sustaining local communities instead.'

'Local people know best how to manage the islands.'

'The reality is we depend on central funding and so at some point there has to be centralised decision no matter how much we want local decisions.'

'Strategic funding decisions on transport must be central, but action implemented through local decision-making.'

'Central government does not have the knowledge of local situations.'

'Since central government is unlikely to be anticipatory, it may be up to us to look after ourselves anyway.'

Source: Adapted from Brown et al. 2005

---

ecological and cultural change. Thus, Stephanie Rotarangi and Darryn Russell (2009: 209) claim that 'to be indigenous is to be resilient' and indeed that 'the maintenance and evolution of identity and culture of indigenous people and communities is premised on such resilience'. In this view, the very fact that indigenous peoples persist in the modern age is testament to their resilience. Yet there is also a strong narrative in the scientific and policy literature on how indigenous people may be especially vulnerable to changes and shocks. Does being indigenous mean people are more resilient, or more vulnerable, or both? I look at some of the studies around resilience 'at the margins' (meaning marginalised and peripheral to the mainstream geographically, politically, culturally and often economically) to understand these and other apparent contradictions. My insights also stem from participation in the workshop which Rotarangi and Russell base their paper on.

There is no one agreed definition of indigenous people. Self-identification as indigenous or tribal is considered a fundamental criterion by United Nations and other agencies. Generally one or more of the following factors are considered as characteristics of indigenous people (from *United Nations Development Group guidelines on indigenous peoples' issues*, February 2008, p8):

- Occupation and use of specified territory.
- Cultural distinctiveness including aspects of language, social organisation, religious and spiritual values, modes of production, laws and institutions.
- Self-identification as well as recognition by other groups, or by state authorities, as a distinct collectivity.
- Experience of subjugation, marginalisation, dispossession, exclusion or discrimination (whether or not these conditions persist).

Each and all of these factors might shape the resilience and vulnerability of indigenous people and influence how indigenous people respond to different changes. Indigenous societies tend to operate at small or local scales often in remote localities, with a strong sense of place. Thus indigenous societies have historical links to land and natural resources, and place-based knowledge that informs their resource management. Governance and management systems often draw on oral traditional, local and traditional knowledge, and connection to ancestral places. But indigenousness also raises important questions about what changes and what stays the same in communities and cultures, bringing to the fore the tensions between adapting and maintaining identity that are at the heart of resilience. For example, David Kepue Ole Nkedianya's research (2010) asks this critical question: how far can the Maasai change but still be Maasai? When livelihoods, lifestyles and place all change, then how is identity, culture and resilience maintained?

McCubbin and McCubbin (2005) observe the interplay between vulnerability and resilience in their study of indigenous Hawaiian families. Indigenous people are more at risk from disease and mortality because of their marginalisation and exposure, but shared identity and culture play an important role in resilience. In particular a family sense of coherence – including comprehensibility, manageability and meaningfulness – constructs shared and holistic views of well-being and wholeness. Cultural practices, such as storytelling and 'self-talk', might also contribute, and there is a strong sense of resilience as relational. So despite societal construction of vulnerability, there also exists a unique set of values and practices which build resilience. Indeed, much writing emphasises how indigenous peoples' worldviews and knowledge may be at odds with conventional scientific constructions of humans and nature. Rotarangi and Russell (2009) observe that many indigenous societies see the ecological realm as not merely connected but inseparable from the economic, the spiritual, the personal or the collective, constructing a very different conception of a social ecological system.

A key question, and one raised by Rotarangi and Russell's paper, is the extent to which traditional practices of particular groups have been able to enhance resilience of communities to ecological shocks or have stifled resilience through rigid rules and practice. Using case studies shown in Table 4.3, Rotarangi and Russell highlight the tensions or dialectical relationship between staying the same and adapting; between who controls change and who defines responses and who has the power and authority to shape change and take anticipatory action.

Highlighting the contribution of ethnographic analysis, Rival (2009) sees resilience as being materialised and enacted through what she calls 'indigenous intelligence', which is related to indigenous or local knowledge. Indigenous intelligence fundamentally depends on learning how to live in environments and is inherently dynamic. The two examples Rival highlights in her analysis demonstrate this. In examining water management and landscape restoration her analysis shows how ecological insights and social innovation develop place-based strategies. This illustrates the power and capacity of indigenous people to build resilience through collective action. Conversely, she also describes the erosion and devaluation of indigenous knowledge and practices – a more conventional narrative of indigenous knowledge – among livelihood and agriculture systems of Makushi farmers in the savannah region in Guyana.

Salick and Ross (2009) summarise some of the various arguments concerning indigenous people, indigenous knowledge and climate change. These include the view of indigenous people as passive and helpless in the face of change and of them as confined to marginal environments – which are often especially threatened by contemporary climate change. Conversely the study of traditional and indigenous people shows them to be highly adaptive and that changes in climate have often been the spur for innovation

Table 4.3 Indigenous people and change: case studies

Case	Ecological trigger	Result	Cultural response
Alaska	Rapid climate change	Change in availability of traditional food	Subsistence communities change resource sharing
Central Australia	Weed invasion and changing climate	Loss of traditional ecosystems	Corresponding loss of language and lifestyle
Seri fishers, Mexico	Fishery overexploitation	Loss of traditional employment	Ethnographic and participatory approaches
New Zealand Maori	Deforestation	Loss of traditional subsistence	Unprecedented combining of resources between families to ensure cultural values not lost
Kenyan Maasiland	Seasonally mobile farmers confined to designated reserves	Change in traditional grazing methods	Animals become commodity as opposed to a mechanism of social connectedness
South African amaXhosa	Grazing limits	Increase in woody weeds results in more people living closer together	Interplay of beliefs and changing livelihoods

Source: Rotarangi and Russell 2009: 210

and 'people have faced climate change and adapted to it since our species evolved' (Salick and Ross 2009: 1); the invention of agriculture was one such adaptation. However, in order to adapt, indigenous people require a diversity of crops and varieties, and of environments. Salick and Ross and others argue that indigenous knowledge is critical for observations of climate change and other changes – and to identify possible adaptations. Indigenous knowledge thus plays a key and active role in resilience of indigenous peoples – a view which concurs with Rival's (2009) conceptualisation of indigenous intelligence and Berkes' (2007) account of traditional ecological knowledge and adaptive management. Furthermore, diverse and alternative worldviews open up spaces for innovation and possibilities to reassess relationships between humans and nature. In this vein, Glen David Kuecker and Thomas D. Hall (2011) understand these processes as core-periphery relationships, and propose marginalised societies (such as indigenous people) as being more emergent and thus more resilient.

Similarly Turner et al. (2003) posit that indigenous people who live in territories which traverse ecological and cultural 'edges' have greater capacity for flexibility and hence greater resilience. Not only do ecological and cultural edges increase resilience because they expand the diversity of resources people can draw on to support livelihoods, they also enhance the biological and cultural diversity of a landscape and allow for the exchange of oral histories, technologies, information and goods necessary to adapt to both expected and unexpected changes in ecological and social systems. The authors furthermore suggest that people create these edges as sites for the transfer and reproduction of place, or culturally specific knowledge necessary for adaptation.

These studies posit that indigenous peoples' sense of place and identity, cultural beliefs and worldviews, in addition to specialist knowledge might characterise resilience; their survival attests to their persistence. Yet their vulnerability might be related to the ecological, as well as economic and cultural marginality. Their resilience again is often articulated as resistance and rootedness, and the persistence of identity in the face of multiple stressors.

Everyday forms of resilience

Using a political ecology lens to understand the lived experiences of resilience provides some very diverse perspectives as well as insights into the social constructions of resilience. The vignettes I've used here provide very different contexts and very different views on multi-layered, multi-scale and multi-faceted resilience. Yet they each inform a socially situated understanding of resilience, and help to develop a view of resilience which has greater traction for development and environmental change. I refer to this as everyday forms of resilience; the term is inspired by James Scott's work and particularly the title of his book, *Weapons of the weak: Everyday forms of peasant resistance* (1987). But it also resonates strongly with Ann Masten's (2001) conceptualisation of resilience as 'ordinary magic' that I discuss in Chapter 3. Everyday forms of resilience represent the strategies and struggles of people dealing with change, and a more agency-centred view of resilience. This concept aims to meld the theoretical and conceptual lessons learned from both social ecological systems and human development fields, with a grounded and situated analysis of peoples' lived experiences of change. Here I summarise five key dimensions of resilience that the empirical insights add to or amplify in conventional interdisciplinary perspectives on resilience, and which are key components of everyday forms of resilience.

Power asymmetries and resistance

Power underscores the ability of people and communities, institutions and social ecological systems to deal with change. The vignettes reveal some of the ways in which power dynamics and asymmetries play out; in struggles about access to land and resources between men and women – sometimes in the same households – in Cameroon; and between upstream and downstream water users in Mozambique. Power manifests in the differential capacities and therefore the different forms for resilience which people can draw on, but it also affects how the processes of change occur. The discussions around resilience and indigenous people highlight some of the issues around rights, self-determination and justice associated with negotiating and determining change. There have been many observations, covered in earlier chapters, that resilience studies are apolitical. Fabinyi et al. (2014), for example, assert that the main way in which institutions and governance issues in particular are addressed are ahistorical, apolitical and highly normative.

A wide literature addresses these issues at different scales, as demonstrated by the vignettes here, presenting more differentiated views of power dynamics within and between societies. For example, Terry (2011) adopts a human security approach to analysing the interwoven stressors experienced by villagers in rural Uganda. Her findings highlight how social conflicts in the form of everyday contestations and tensions, including the abuse of power by state representatives at a local level, are important drivers of human insecurity and are differentiated in many important ways. The threats associated with climate change or long-term environmental change are interlinked and experienced through social stressors such as the intra-household power disparities, and the pressures of bride-price and other social obligations, as well as the dysfunctional operation of health and education infrastructure, and chronic ill health include high HIV prevalence rates.

The active resistance to external policy interventions designed to build resilience is also documented. For example, in New Zealand, Hayward (2008) reports how priority setting over coastal planning is fraught, as is the case in many parts of the UK. Hayward explains coastal community rejection of managed retreat policies in New Zealand as a result of polarisation of local views, and a division between private rights and wider public goods. In New Zealand coasts are important sites of economic activity but also have deeply engrained cultural values and heritage. Coasts are often significant sacred cultural sites for New Zealand's indigenous community, and are described as 'intrinsic to Maori identity' and 'fundamental to all aspects of Maori wellbeing' (Hayward 2008: 48). A policy of managed retreat threatened sacred sites and holiday homes, but local decision-making seemed unable to reconcile competing interests and voices, partly because of lack of resources, and poor consultation procedures. Hayward reports how the rich and powerful (the articulate, second-home owning elite) have disproportionate influence and say in decision-making, challenging the fairness of planning processes and outcomes. Thus she warns that, 'without careful thought, local input into decision-making about global problems like climate change risks becoming a futile, isolationist and unjust exercise which advantages well resourced property owners while exhausting wider community reserves' (Haywood 2008: 57).

Cross-scale interactions and interventions

The vignettes reinforce the need to look at multiple changes and especially how different changes interact, and how risks are amplified by on-going stresses. Women's experience of change in Cameroon, driven by regional and global economic change and structural adjustment policies and mediated by social institutions which determine property rights and access to forests, and intra-household power asymmetries and decision-making, demonstrate the cross-scale and multiple-stressor interactions that Leichenko and O'Brien (2008) documented in India as 'double exposures'.

So the synergistic impacts of external 'drivers' and international social dynamics often produce the enhanced impact, and it is this phenomenon that a resilience approach must address if it is to inform problems of chronic and intransigent poverty – this is further explored in Chapter 6. It shows how the criticisms of social ecological system resilience as over-emphasising exogenous drivers are justified, but also that there needs to be understanding of dynamic feedbacks and synergies across scales.

A critical observation is on the role of external policy interventions, and how these might conspire and may even undermine people's resilience. We posited this as a form of maladaptation or 'misfit' in our analysis of the resilience and vulnerability of coastal communities in Mozambique and Tanzania. But even in the narratives of Orkney islanders, in a developed country, the suspicion that external policies – the subsidies for transport, and reliance on goods imported from the mainland – was undermining self-sufficiency and resilience was evident.

This supports Mearns and Norton's (2009: 3) observation that, 'in the short term, the biggest impact on poor people may result less from changing climate itself than from policies adopted to mitigate climate change'. This issue is further discussed in Chapter 5, which demonstrates how a resilience approach might provide analytical insights into these 'maladaptations' (Barnett & O'Neill 2010: 3) and might help to formulate policies which avoid them.

Social dynamics of resilience

The vignettes highlight the 'winners and losers' and the socially differentiated nature of resilience. The HIV/AIDS pandemic in Uganda affected different people in different ways; the elderly took a burden of care and were left without care themselves as they aged, and the life opportunities of the young were devastated. Men and women in Cameroon found different costs and opportunities in the financial crisis and structural adjustment policies. How does this disaggregated view map onto the systems view taken in social ecological resilience literature? Fabinyi et al. (2014) detail how and why the analytical lens conventionally applied by resilience scholars emphasises consensus and homogeneity over contestation and difference, and – despite a recent social turn in resilience studies and focus on community resilience and social ecological systems – still underplays the role of social difference (Brown 2014) as discussed in Chapter 3. But the human development literature shows explicitly why resilience is socially differentiated. Bringing together these fields informed by a socially situated perspective on experiential resilience, allows a more complex but socially informed view. It reveals resilience to be the site of social relations, struggles and dynamics, intricately connected to how societies work. It shows how people juggle multiple stresses and multiple demands and how they use and access resources at hand, mediated by various political and economic factors and relations.

Contested knowledges and values

Gómez-Baggethun et al. (2013) introducing a special issue of *Ecology and Society* on 'Traditional ecological knowledge and global environmental change' identify three important links between traditional ecological knowledge and resilience. First, they posit traditional ecological knowledge as itself resilient. Despite worldwide erosion of traditional knowledge, important pockets – or 'refugia' – persist amongst communities in both developing and developed countries. Although specific components or aspects of knowledge may be lost, and a process of hybridisation with other knowledge forms occurs, what the authors identify as critical, is whether society retains the ability to generate, transform, transmit and apply traditional ecological knowledge – in other words the system of knowledge endures. This recognises the dynamic and adaptive nature of traditional ecological knowledge. They comment that maintaining the capacity for an evolving traditional ecological knowledge requires securing rights of indigenous and other marginalised or traditional peoples to control and develop their knowledge; attempts to 'preserve' traditional ecological knowledge in fossilised forms are certain to fail.

Second, traditional ecological knowledge is increasingly acknowledged as a source of resilience, in retaining valuable biological and cultural information. These knowledge systems and the practices and institutions they encompass, are recognised for their contribution to sustaining biodiversity and ecosystem services. This knowledge has co-evolved from long-term observation of ecological dynamics and experimentation, learning from crises and mistakes, representing important reservoirs for adaptive management on complex social ecological systems.

Third, traditional ecological knowledge often provides insights into change and gives very different perspectives on the stressors that affect peoples' lives and livelihoods, and local social ecological systems. This is demonstrated in the examples of mental models developed with communities in coastal Mozambique and Tanzania (see Figures 4.2 and 4.3); the stressors people identified as affecting them, and the priority they gave to different factors reveals quite different perspectives to the priorities of development and environmental organisations. Local representations of environmental change based on traditional ecological knowledge, attuned to local notions of values and ecological dynamics, as well as local systems for representing, monitoring and understanding change are critical for development of appropriate and meaningful – and supported – adaptation strategies.

Despite observations of hybridisation (Blaikie et al. 1997) or integration (Bohensky & Maru 2011), or even Rival's (2009) indigenous intelligence, and recognition that indigenous knowledge comprises more than static, taxonomic components but a body of knowledge, beliefs, traditions, practices, institutions, and worldviews, knowledge contestations are intensely political. The vignette on indigenous and marginalised people highlights the contestations around both power and knowledge around resilience, and the significance of different worldviews, values and conceptualisations of people–nature relations. This will influence the framing of resilience and resource management.

Satterfield et al. (2013) discuss these issues in the context of recent innovations by resource planners to 'recognise that indigenous knowledge and the cultural knowledge they hold as key to good environmental management' (p103). They highlight recent national legislation in Bolivia and Ecuador that grants rights to nature, strongly influenced by indigenous views of human-environment, and providing frameworks and opportunities to negotiate resource management conflicts and contestations. They provide three illuminating examples of how these negotiations play out in environmental assessment where practitioners seek to better integrate indigenous views and values into decision-making and planning procedures. In western Canada, river flows affected by dams were the subject of a set of discussions to support the development of an adaptive management strategy. The lower Bridge River runs through the traditional territory of the St'at'imc First Nation. This group have a strong sense of environmental stewardship, extending to shared sense of responsibility for the river which was seen as intrinsic to St'at'imc identity and culture. The river has spiritual values and 'voice' quite beyond the remit of conventional analysis. In this way, indigenous or traditional knowledge becomes more than a resource, or a means of mobilising support for resource management (for example around sacred sites or keystone species), but about opening negotiation around a different set of values, beliefs and practices, beyond the instrumental. This then presents the opportunity for a pluralist perspective on learning and adapting to change, and the possibility of new emergent knowledge systems.

Situated resilience and sense of place

In each of the vignettes there is a strong sense of resilience as situated in a specific place; resilience reflects a rootedness that is not just about place, but encompasses identity and belonging. This is articulated especially strongly in the cases from Orkney and for indigenous people. I discuss in the following chapter how attachment to place and occupation may affect capacity to adapt to change, but the situated nature of resilience is important from a political ecology perspective because it anchors our understanding of resilience to a set of social practices and relations, and an ecology of place. Sense of place and place attachment have been researched extensively by psychologists and human geographers, but they are slowly finding a role in analysis of adaptive capacity and adaptation to environmental change. In Lucy Faulkner's research (2014), sense of place was found to be a critical precursor for community resilience. But the sense of place is rooted not just in physical space, but amongst community and relations.

Christopher Lyon's discussion (2014) presents a novel framework for understanding place in the context of resilience which emphasises both physical dimensions (which he refers to as 'incarnate place'), the unconscious social and physical manifestations of heritage and culture ('discarnate place'); and place attachment and the emotive element people articulate in their relationship to place ('chimerical place'). This makes clear the link between the physical and the emotional aspects of place, extending sense of place as conventionally studied. I develop these notions more in later chapters and particularly in relation to the importance of rootedness within situated resilience. Raymond et al. (2010) have also explored these different dimensions, and distinguish five components to place attachment: place identity, place dependence, nature bonding, family bonding and friend bonding, again combining functional, emotional and social connections to place. Each of these aspects is evident in the vignettes presented here, and provides an important dimension of resilience which is relatively underrepresented in social ecological systems and human development literatures.

Resistance, rootedness and resourcefulness

The concept of everyday forms of resilience builds on exploration of experiential resilience with the insights distilled across the fields reviewed in Chapter 3. This earlier discussion highlights the attempts to link the social and the ecological sub-systems in the social ecological systems approach. It shows how human development perspectives, initially focused on the traits and characteristics of individuals, evolved to develop an 'ecological' perspective, situating an individual within a set of relationships and transactions in society and place. The narratives and perspectives from the vignettes here demonstrate strongly articulated sense of place, community and identity that contributes to peoples' perception and experience of resilience. I synthesise three core themes from this analysis that I use to shape and re-vision resilience for development. These are resistance, to encompass the contested, politicised nature of peoples struggle to respond to, negotiate and mobilise for change; rootedness, to accentuate the importance of place and the situated nature of resilience in highly dynamic contexts; and resourcefulness, to emphasise the capacities and agency of different social actors and their social ecological system to manage and shape change in both positive and negative ways. But in strengthening the social perspective on resilience, I also acknowledge the necessity to better integrate the ecological as both the site for struggles and actions around dealing with change, but also as an active agent in change.

Note

1   A **vignette** is used in literature to mean a short impressionistic scene which gives a particular insight into a character, idea or setting. I summarise studies here to highlight aspects of resilience hitherto under-emphasised in literature and which are of particular importance to international development. I've been directly or indirectly involved (through colleagues) in most of these cases.

References

Adger WN, Paavola J, Huq S and Mace MJ (eds) (2006) *Fairness in adaptation to climate change*. Cambridge, MA: MIT Press.

Almedom A and Tumwine J (2008) Resilience to disasters: A paradigm shift from vulnerability to strength. *African Health Sciences*, 8(5), 1–5.

Barnett J and O'Neill S (2010) Maladaptation. *Global Environmental Change*, 20(2), 211–13.

Berkes F (2007) Understanding uncertainty and reducing vulnerability: Lessons from resilience thinking. *Natural Hazards*, 41(2), 283–95.

Blaikie P, Cannon T, Davis I and Wisner B (1994) *At risk: Natural hazards, people's vulnerability and disasters*. London: Routledge.

Blaikie P, Brown K, Stocking M, Tang L, Dixon P and Sillitoe P (1997) Knowledge in action: Local knowledge as a development resource and barriers to its incorporation in natural resource research and development. *Agricultural Systems*, 55(2), 217–37.

Bohensky E and Maru Y (2011) Indigenous knowledge, science, and resilience: What have we learned from a decade of international literature on 'integration'. *Ecology and Society*, 16(4), 6.

Brouwer R and Nhassengo J (2006) About bridges and bonds: Community responses to the 2000 floods in Mabalane district, Mozambique. *Disasters*, 30(2), 234–55.

Brown K (2014) Global environmental change I: A social turn for resilience? *Progress in Human Geography*, 38(1), 107–17.

Brown K and Ekoko F (2001) Forest encounters: Synergy among agents of forest change in Southern Cameroon. *Society & Natural Resources*, 14(4), 269–90.

Brown K and Lapuyade S (2001) A livelihood from the forest: Gendered visions of social, economic and environmental change in Southern Cameroon. *Journal of International Development*, 13(8), 1131–49.

Brown K, Few R, Tompkins EL, Tsimplis M and Sortti T (2005) *Responding to climate change: Inclusive and integrated coastal analysis*. Norwich, Available from: http://tyndall.ac.uk/sites/default/files/t2_42.pdf (accessed 12 August 2015).

Bunce M, Rosendo S and Brown K (2010a) Perceptions of climate change, multiple stressors and livelihoods on marginal African coasts. *Environment, Development and Sustainability*, 12(3), 407–40.

Bunce M, Brown K and Rosendo S (2010b) Policy misfits, climate change and cross-scale vulnerability in coastal Africa: How development projects undermine resilience. *Environmental Science and Policy*, 13(6), 485–97.

Eriksen S and Lind J (2009) Adaptation as a political process: Adjusting to drought and conflict in Kenya's drylands. *Environmental Management*, 43(5), 817–35.

Eriksen S and Silva J (2009) The vulnerability context of a savanna area in Mozambique: Household drought coping strategies and responses to economic change. *Environmental Science & Policy*, 12(1), 33–52.

Fabinyi M, Evans L and Foale S (2014) Social ecological systems, social diversity, and power: Insights from anthropology and political ecology. *Ecology and Society*, 19(4), 28.

Faulkner LC (2014) Assessing community resilience in North Corwall: Local perceptions into responding to changing risk landscapes. MRes dissertation. University of Exeter.

Few R, Brown K and Tompkins EL (2007a) Climate change and coastal management decisions: Insights from Christchurch Bay, UK. *Coastal Management*, 35(2–3), 255–70.

Few R, Brown K and Tompkins EL (2007b) Public participation and climate change adaptation: Avoiding the illusion of inclusion. *Climate Policy*, 7(1), 46–59.

Folke C (2006) Resilience: The emergence of a perspective for social–ecological systems analyses. *Global Environmental Change*, 16(3), 253–67.

Gómez-Baggethun E, Corbera E and Reyes-García V (2013) Traditional ecological knowledge and global environmental change: Research findings and policy implications. *Ecology and Society*, 18(4), 72.

Guyer J (1984) Family and farm in southern Cameroon. Boston University, African Studies Center.

Hayward B (2008) 'Nowhere far from the sea': Political challenges of coastal adaptation to climate change in New Zealand. *Political Science*, 60(1), 47–59.

Hayward B (2013) Rethinking resilience: Reflections on the earthquakes in Christchurch, New Zealand, 2010 and 2011. *Ecology and Society*, 18(4), 37.

Kuecker GD and Hall TD (2011) Resilience and community in the age of world-system collapse. *Nature and Culture*, 6(1), 18–40.

Leichenko R and O'Brien K (2008) *Environmental change and globalisation: Double exposures*. Oxford: Oxford University Press.

Lyon C (2014) Place systems and social resilience: A framework for understanding place in social adaptation, resilience, and transformation. *Society & Natural Resources*, 27(10), 1009–23.

McCubbin L and McCubbin H (2005) Culture and ethnic identity in family resilience: Dynamic processes in trauma and transformation of indigenous people. In: Ungar M (ed.), *Handbook for working with children and youth: Pathways to resilience across cultures and contexts*, Thousand Oaks, CA/London/New Delhi: SAGE Publications, 27–44.

Masten A (2001) Ordinary magic: Resilience processes in development. *American Psychologist*, 56(3), 227–38.

Mearns R and Norton A (2009) *Social dimensions of climate change: Equity and vulnerability in a warming world*. Washington, DC: World Bank Publications.

Miller F, Osbahr H and Boyd E (2010) Resilience and vulnerability: Complementary or conflicting concepts. *Ecology and Society*, 15(3), 11.

Nelson D, Adger WN and Brown K (2007) Adaptation to environmental change: Contributions of a resilience framework. *Annual Review of Environment and Resources*, 32(1), 395–419.

Nkedianya DKO (2010) The Maasai of East Africa: Demographic characteristics, drought coping strategies, and livestock-wealth inequalities. PhD thesis. University of Edinburgh.

O'Brien K and Leichenko RM (2003) Winners and losers in the context of global change. *Annals of the Association of American Geographers*, 93(1), 89–103.

Raymond C, Brown G and Weber D (2010) The measurement of place attachment: Personal, community, and environmental connections. *Journal of Environmental Psychology*, 30(4), 422–34.

Ribot J (2009) Vulnerability does not fall from the sky: Towards multi-scale, pro-poor climate policy. In: Mearns R and Norton A (eds), *Social dimensions of climate change: Equity and vulnerability in a warming world*, Washington, DC: World Bank Publications, p. 319.

Rival L (2009) The resilience of indigenous intelligence. In: Hastrup K (ed.), *The question of resilience: Social responses to climate change*, Copenhagen: The Royal Danish Academy of Sciences and Letters, 293–313.

Rotarangi S and Russell D (2009) Social ecological resilience thinking: Can indigenous culture guide environmental management? *Journal of the Royal Society of New Zealand*, 39(4), 209–13.

Salick J and Ross N (2009) Traditional peoples and climate change. *Global Environmental Change*, 19(2), 137–39.

Satterfield T, Gregory R and Klain S (2013) Culture, intangibles and metrics in environmental management. *Journal of Environmental Management*, 117, 103–114.

Scott J (1987) *Weapons of the weak: Everyday forms of peasant resistance.* Yale University Press, 389.

Seeley J, Wolff B, Kabunga E, Tumwekwase G and Grosskurth H (2009) 'This is where we buried our sons': People of advanced old age coping with the impact of the AIDS epidemic in a resource-poor setting in rural Uganda. *Ageing and Society*, 29(1), 115–34.

Seeley J, Dercon S and Barnett T (2010) The effects of HIV/AIDS on rural communities in East Africa: A 20-year perspective. *Tropical Medicine and International Health*, 15(3), 329–35.

Smucker TA and Wisner B (2008) Changing household responses to drought in Tharaka, Kenya: Vulnerability, persistence and challenge. *Disasters*, 32(2), 190–215.

Sunderlin W, Ndoye O, Bikie H, Laporte N, Mertens B and Pokam J (2000) Economic crisis, small-scale agriculture, and forest cover change in southern Cameroon. *Environmental Conservation*, 27(3), 284–90.

Terry G (2011) Climate, change and insecurity: Views from a Gisu hillside. PhD thesis. School of International Development, University of East Anglia.

Tompkins EL, Few R and Brown K (2008) Scenario-based stakeholder engagement: Incorporating stakeholders preferences into coastal planning for climate change. *Journal of Environmental Management*, 88(4), 1580–92.

Turner N, Davidson-Hunt I and O'Flaherty M (2003) Living on the edge: Ecological and cultural edges as sources of diversity for social—ecological resilience. *Human Ecology*, 31(3), 439–61.

United Nations (2008) *United Nations Development Group guidelines on indigenous peoples' issues.* Available from: www2.ohchr.org/english/issues/indigenous/docs/guidelines.pdf (accessed 12 August 2015).

5   Adaptation in a changing climate

This chapter outlines how a socially informed and pluralist perspective on resilience can enhance understanding of societal responses to disturbances and shocks. It relates to current science and policy on climate change adaptation and development. This continues to build my case for a re-visioning of resilience that can contribute to analysis of the dynamics of change in society, especially in context of global environmental change and development. So far I've looked at the framing of resilience, the discourses constructed around it in policy arenas, its meaning in different scientific fields, and its empirical experiences. I now focus more specifically on climate change adaptation and how it has become an important concern for international development. This chapter examines how a resilience perspective provides new insights for knowledge and policy, and supports more transformative responses to environmental change. In particular it reflects on current approaches to climate change adaptation and how a multi-faceted resilience lens brings insights into socially differentiated capacities to deal with change at multiple scales.

Adaptation and development: adaptation as development

Climate change adaptation has become a core issue in international development. Chapter 2 shows how development policy, particularly concerned with climate change and environmental management, has increasingly emphasised and used resilience concepts and language. This is closely related to heightened focus on climate change and adaptation. For example, Box 5.1 shows how major funding streams have developed, demonstrating large-scale and widespread strategic interest in adaptation. Increasingly, development agencies – including multinational donors, bilateral agencies and NGOs – recognise the need for adaptation to climate change, and frequently advocate for mainstreaming climate change adaptation into development policy.

Box 5.1 International funding for climate change adaption

The international climate change regime has long recognised the need for adapting to the now observed and inevitable future impacts of climate change. Countries have discussed how adaptation should be prioritised and funded since the Conferences of the Parties of UNFCCC meetings from Delhi in 2001.

The Adaptation Fund was initiated to 'finance concrete adaptation projects and programmes'. The majority of efforts and funds under the UNFCCC had, by contrast, been focused on helping developing countries move away from fossil fuel dependence and reduce carbon loss associated with deforestation and land use change. Funds to help nation states adapt to the impacts of climate change have been more recent. The tensions between international development assistance and climate change related funding have always been at the forefront of discussions.

International climate negotiations have resulted in a number of planning initiatives and funds related to adaptation, including the Nairobi Workplan on Adaptation to co-ordinate best practice, the Adaptation Fund, Least Developed Country Fund and National Adaptation Plans of Action (NAPAs) undertaken by most of the world's least developed countries over the past decade.

The Copenhagen Agreement in 2009 accelerated aspirations for climate change funding. Both the Copenhagen Agreement and the subsequent Cancun Accords of November 2010 promoted a Green Climate Fund with an intention to raise $100 billion by 2020 and to spend this on climate change through transforming energy systems, reducing deforestation and promoting adaptation.

The first two years of the Green Climate Fund, known as the Fast Start period, have seen pledges of almost $35 billion for climate change. Almost 70 per cent of these funds have been focused on energy and mitigation projects. Around $5 billion was pledged for adaptation assistance in the Fast Start period.

There are significant challenges associated with flows of international finance for adaptation, summarised by Fransen et al. (2013):

- Governance must be transparent, accountable and give voice and representation to vulnerable populations over and beyond representing national government interests.
- Funding should not substitute for international development transfers.

- There is a risk of simply viewing problems of development through a climate lens. Adaptation interventions need to build generic adaptive capacity, ranging from education to community empowerment.
- There are significant challenges in building resilience in the face of uncertainty about how severely the impacts of climate change may be experienced in particular systems and localities.
- There are challenges in promoting sustainable and resilient adaptations and hence avoiding adaptations that make others vulnerable across space and time.

Source: Author's own

Yet this focus on climate change and adaptation in the development community is a relatively recent phenomenon. As Bassett and Fogelman (2013) note, ideas about adaptation came relatively late in the climate change discussions, first really emphasised in the Intergovernmental Panel on Climate Change's (IPCC's) *Third Assessment Report* in 2001. Bassett and Fogelman attribute the rise of adaptation in climate change science and policy as prompted by first, a recognition that climate change was already happening and impacts were already being experienced around the world; and second, the apparent failure of climate change mitigation policy and action. Yet incursions of climate change and adaptation into mainstream international development came later. A key turning point for the international development community was in 2006 at the negotiations at the annual Conference of the Parties to the UN Framework Convention on Climate Change (UNFCCC) in Marrakesh and the launch of the Climate Adaptation Fund. This galvanised international development agencies, NGOs and developing country organisations to look for ways and means to operationalise and often to 'mainstream' climate change adaptation within development planning, investments and policies.

The Tyndall Centre convened a conference in 2001 bringing together climate change and development scholars and practitioners to prompt discussion on how climate change was emerging as an important issue for international development, and especially to examine how the incipient adaptation agenda might affect development. Surprisingly, at that time, less than a decade and a half ago as of this writing, climate change and adaptation remained outside the core business and central concerns of development. We followed this agenda-setting conference with a paper published in *Progress in Development Studies* (Adger et al. 2003), making the case for development to engage with climate change adaptation.

We highlighted an apparent paradox; international climate change negotiations recognised the need for adaptation, but identified the primary means of addressing this was through transfer of technology, know-how and funds to governments, whereas poor people in poor countries were already experiencing and adapting to climate variability, unsupported by governments and development agencies in most instances. This second aspect was highlighted by participants and discussions at the meeting. Yet these two perspectives typified the discourses and approaches of, on the one hand, states and climate change scientists participating in discussions at a multilateral level, and on the other, the emphasis of development practitioners and NGOs focused on poverty alleviation in developing countries. This paradox throws up many critical issues for climate change adaptation and development which have been argued in many political and scientific fora since 2001; for example, around justice and equity, uncertainty and prediction, and about the very nature and potential role of adaptation – who adapts and to what, and about when, where and how adaptation happens.

Jessica Ayers' (2010) research centres on the same paradox, and examines how adaptation policy is applied and how it plays out in the state apparatus and project context of developing countries, using case studies of Bangladesh and Nepal. She asks how and whether adaptation, driven by international priorities and resourced through international climate funds, can reflect local practices and priorities and be locally inclusive. In examining the relationship between adaptation actions and development practice and policy, Ayers' analysis identifies three different types of adaptation, shown in Table 5.1. These illustrate how the relationship between adaptation and development is understood. The first approach targets intervention to address specific climate change hazards or impacts – it doesn't question whether development priorities or strategies need to change. This approach is driven chiefly by climate change policy agendas and actors, and gives emphasis to the countries and communities most exposed to climate hazards. The second approach recognises that current development should change to take account of climate change in order to reduce the most harmful impacts. Again this is a largely climate change-driven agenda, seeking to make current development strategies more 'resilient' to climate change. The discussion in Chapter 2 highlights these interventions and the dominant framing and discourses around them – of 'adapting' current development in face of climate change, and of making development more resilient in face of climate change. The third approach Ayers identifies views adaptation primarily as development, and emphasises building adaptive capacity as the means of ameliorating vulnerability not only to climate change impacts, but to other stressors too. In Ayers' analysis only development organisations are promoting and implementing this approach to adaptation. But even in this approach, the assumption that adaptation is about doing 'development' as usual, and that building capacity to deal with climate change risks is centred on alleviating poverty is prevalent. As this chapter discusses further, this approach lacks consideration of the complex and dynamic nature of risks,

Table 5.1 Approaches to adaptation and development

	Targeted adaptation interventions	Adaptation 'plus' development: 'Climate proofing', 'Climate resilient development'	Adaptation 'as' development: Development as usual
Focus of adaptation	Adaptation addresses the impacts of climate change	Adaptation reduces vulnerability to climate change and variability	Adaptation increases the capacity of people to adapt to climate change and other stresses
Assumptions about vulnerability	Vulnerability is a consequence of exposure to climate change hazards	Vulnerability is a consequence of exposure, sensitivity and adaptive capacity	Vulnerability is directly dependent on adaptive capacity which in turn is determined by development factors
Target population	Countries and communities most exposed to climate change impacts	The climate-vulnerable poor in countries and regions exposed to climate change impacts	The poorest and most marginalised in developing countries
Main actors promoting the approach	UNFCCC, some IPCC actors, some donors	IPCC, some donors	Development NGOs

Source: Adapted from Ayers 2010

their implications over time and for different groups and contexts, and sees vulnerability in relatively simplistic terms, often simply as the poor being most vulnerable. Thus it targets the poorest and marginalised in poor countries in line with conventional development strategies.

Ayers' analysis is informed by empirical examination of policy development and implementation, studying the processes and actions in adaptation policy development. In another analysis, Bassett and Fogelman (2013) undertake an extensive review of recent writings on climate change, scrutinising more than 550 articles published in four leading environmental change journals. They classify these articles according to the approach taken to adaptation, identifying three categories: 'adjustment adaptation', 'reformist adaptation' and 'transformational adaptation'. They find that 70 per cent of the papers they reviewed conformed to the adjustment adaptation category. This suggests more narrowly-defined and technical interventions to address climate change hazards, in line with Ayers' targeted interventions approach, and that entertain no need to challenge or to change strategies. Only three per cent of the papers addressed transformational adaptation, which suggests radical change is needed to adapt to climate change. Some 27 per cent of papers were categorised as focusing on reformist adaptation, recognising that some kind of systemic change is required and that interventions should go beyond a 'technical fix'. Reformist adaptation suggests that successful adaptation requires changes to institutions and rules, such as inter-sectoral co-ordination and better inclusion of local knowledge and institutions, but seeks to alter the rules rather than fundamentally change the system. Bassett and Fogelman assert that contemporary debates around climate change adaptation are dominated by this IPCC 'impact-led' approach, focused on adaptation as an endpoint or outcome, giving weight to rational and technocratic measures.

In tracing the evolution of these debates around adaptation and linking them to earlier literatures around vulnerability and hazards, Bassett and Fogelman's (2013) analysis gives an intellectual grounding to what might otherwise be seen as political debates around buzzwords, or terms dismissed as 'malleable' or loose concepts. They root these debates in natural hazards literature of the 1970s and early 1980s (e.g. Burton et al. 1978) that proposed adaptation as 'purposeful adjustment', and its political economy critique (e.g. Watts 1983). The political economy approach emphasised the underlying economic and political forces that drive vulnerability, citing marginalisation and 'under-development' as root causes, and criticised the implicit support of the natural hazards school for the 'status quo', which sought to 'naturalise risk by locating it in the hazard itself' (Bassett & Fogelman 2013: 46). The parallels with current debates around adaptation and resilience are clear. The political economy critique itself evolved into the field of political ecology, so these debates have resonance for my analysis and re-visioning of resilience, recognising the systemic, structural causes of vulnerability, but also highlighting the importance of human agency, individual and collective action, efficacy and resistance.

The next two sections look in more depth at how contemporary debates around adaptation and development attempt to forge and define adaptation more closely aligned with poverty alleviation and sustainable development goals, and specifically at ideas around sustainable adaptation, and an adaptation pathways approach that calls for more dynamic process-orientated understandings of adaptation. This then leads to a more detailed and conceptual discussion of the distinctions between adaptation, vulnerability and resilience, and how a resilience lens can inform adaptation and development strategies and actions.

Making adaptation sustainable

How then can adaptation be made more synergistic with development? 'Sustainable adaptation' has been proposed as a way of addressing some of the problems or paradoxes outlined in the previous section, and to make adaptation strategies more closely aligned with sustainable development.

Developed from a series of panels and papers at the Global Environmental Change and Human Security Conference in 2009, I co-edited a special issue of the journal *Climate and Development* with Siri Eriksen (Eriksen & Brown 2011), which sought to explore notions and applications of sustainable adaptation. Ideas about sustainable adaptation have evolved alongside the realisation that many adaptation actions have consequences for sustainability; in social, political and ecological dimensions. Thus Eriksen and colleagues (2011: 7) assert that 'not every response to climate change is a good one' and observe that little attention has been paid to the consequences of adaptation policies and practices for sustainability. They highlight that adaptation may have socially differentiated impacts, and engender both temporal and spatial trade-offs. This means that adaptation actions might lessen the vulnerability of one group but have negative impacts – for example through livelihood options or security or limiting access to resources – on another, as observed in Mozambique coastal communities discussed in Chapter 4. Adaptation may also have unintended negative consequences on other parts of the environment, for example leading to biodiversity loss. The collection of papers included in the *Climate and Development* special issue document such examples (see Eriksen & Brown 2011).

Debates on climate change adaptation have taken place largely outside the broader discourse on sustainable development according to Eriksen et al. (2011). Although sustainable development is often cited in discussions about climate change – for example in the IPCC and other international policy fora – relatively little attention has been paid to identifying factors that create synergies between adaptation and sustainable development. Furthermore, climate change has been largely constructed as an environmental problem which can be solved by reducing greenhouse gas emissions, rather than a more systemic and fundamental problem of society and unsustainable development pathways, which necessitates shifts in social, cultural, political and ethical dimensions (Adger et al. 2012). Indeed the neglect of these dimensions undermines attempts to implement adaptation; a vicious circle ensues whereby cultural and social aspects are ignored, so the adaptation measures are infeasible and poorly supported.

Eriksen et al. (2011) suggest four principles for sustainable adaptation, summarised in Table 5.2. Sustainable adaptation, they claim, can be distinguished from adaptation in general because it qualifies actions in terms of their impacts on social justice and environmental integrity. They promote sustainable adaptation as a means to address three core problems identified in the vulnerability literature. The first problem relates to spatial and temporal trade-offs, which mean that adaptation measures that benefit one group, place or sector may undermine the security and well-being of another. Second, adaptation may not tackle causes of poverty and inequality that make people vulnerable. Third, sustainable adaptation also needs to help define more sustainable development pathways, including low emissions and low-carbon development, energy supply food production and water

Table 5.2 Key principles for sustainable adaptation

Key principle	Description	Example case study and recommendations
1 Recognise the context for vulnerability, including multiple stressors	Individuals, groups and regions are experiencing multiple stressors, besides environmental change.	*Addressing the vulnerability context of poor communities affected by floods and rainstorms in the city of Ilorin, Nigeria.* In Ilorin, the capital city of Kwara State in Nigeria, multiple stressors generate vulnerability and unless socio-economic dimensions are tackled in combination with infrastructure, climate-related extreme events such as heavy rainstorms and flooding will continue to have effects on livelihoods and long-term vulnerability.
2 Acknowledge that different values and interests affect adaptation outcomes	Values and interests play an important yet seldom discussed role in climate change responses. Recognising potential value conflicts can help to identify how adaptation responses taken by one group may affect the vulnerability context of other groups.	*Including the adaptation interests of vulnerable groups in local government policy in Durban, South Africa.* Durban, exposed to both flooding and coastal erosion, illustrates how important it is to develop institutions that focus on social equity and vulnerability in order to achieve sustainable adaptation. In particular, prioritising the needs of vulnerable groups in both development and climate policy processes is critical.
3 Integrate local knowledge into adaptation responses	Different groups and actors produce different knowledge on adaptation. Integrating local knowledge based on the experience of living in a risky place and of observing the natural environment is essential for sustainable adaptation to climate change.	*Building on local knowledge and capacity in risk reduction in Concepción, Chile.* The community of Aguita de la Perdiz consists of mainly informal and illegal settlements, built on landslide prone areas in Concepción, Chile. In 2005 100 out of 282 houses were partially or completely destroyed by a severe storm, but there were no deaths reported, and only a few injuries. The community made use of both past experience and knowledge about which areas would be most exposed and which people would be hardest hit; they organised a refugee camp, evacuated vulnerable community members and guarded houses against robbers.
4 Consider potential feedbacks between local and global processes	Adaptation responses may directly affect the vulnerability of local populations, but can also influence – or be influenced by – larger-scale processes. Feedbacks and linkages can influence both social justice and environmental integrity over both space and time. This raises questions about the sustainability of many adaptation responses.	*Linking adaptation with mitigation and transformations towards a resilient society in Norway.* In Oslo, Norway, there are plans to construct the country's biggest artificial ice rink to enable people to skate despite warming winter conditions. Current adaptations to deteriorating snow and ice conditions (due to warming conditions) appear to focus on preserving existing activities through 'controlling' local environmental conditions in the short term in the face of changing weather conditions, often in ways that involve increased energy use.

Source: Adapted from Eriksen et al. 2011

management. According to this understanding, sustainable adaptation should therefore be a more dynamic process of identifying and implementing strategies that reduce both vulnerability and greenhouse gas emissions.

The principles themselves then suggest a quite different set of criteria than are implied by the approaches that Ayers (2010) identified, or which Bassett and Fogelman (2013) distilled from the scientific literature. Suggestions for sustainable adaptation constitute a radical change in thinking and planning adaptation, which addresses some of the core issues concerning more dynamic and complex systems approaches. Thus, understanding context and addressing multiple stressors, rather than specific climate change impacts; incorporating plural values, interests and knowledges into design and implementation; and taking account of cross-scale feedbacks add new dimensions to current adaptation, and each build on core concepts from resilience thinking. These are discussed in the following sections, reflecting in the light of different examples of adaptation actions. The principles link to major themes developed in this book. The first principle accentuates the lived experiences or everyday forms of resilience. The second highlights the contested nature of resilience and the possible role of resistance and power struggles in responding to change. The third links adaptation and local knowledge to rootedness. The fourth principle exposes the cross-scale dynamic of adaptation and possible maladaptation across scales.

But the notion of sustainable adaptation is also potentially plagued with difficulties in terms of the loose definitions and malleable concepts associated with 'sustainable development', and although its normative goals are laudable, the practical challenges of implementation are considerable. I have commented earlier (Brown 2011) on how policy makers also make simplistic assumptions about what links poverty and climate change impacts, and this may also undermine the effectiveness of climate change responses. Much existing literature – and policy – is based on the assumption that alleviating poverty will enable people to adapt to climate change, such that for Mearns and Norton (2010: 30) 'development is the best form of adaptation'. In other words, that there is an almost complete overlap between those who are poor and those who are vulnerable to the impacts of climate change. Proponents of the 'Adaptation as Development' approach identified by Ayers (2010; see Table 5.1) adhere to this view. Yet Eriksen and O'Brien (2007) very carefully show why this isn't the case, highlighting how the experiences and states of vulnerability and poverty are distinct. They observe that many of the processes that generate vulnerability to climate change are closely associated with poverty, but that there is no *direct* mapping of climate change vulnerability and poverty. Not all poor people are vulnerable to climate change and certainly not all in the same ways; and not all of those vulnerable to climate change are poor.

Vulnerability is not solely determined by the direct impacts of climate change on resources; people's ability to access resources, the social and political relations and the economic forces which enable or constrain them are critical. Although poverty and marginalisation are key drivers of vulnerability, and constrain individuals and households in their coping and long-term adaptation, vulnerability to future climate change is likely to have distinct new patterns and characteristics. It is highly likely that those people who are currently marginalised will be among the more severely affected, and indeed it has been argued that climate change is likely to exacerbate and reinforce inequalities both through dynamic vulnerability and the processes of adaptation itself (Adger et al. 2003). But new vulnerabilities are likely to emerge, especially as climate change takes multiple and uncertain turns and as it interacts with existing and emerging stressors. These factors all point to why an approach that builds resilience and adaptive capacity, will be more sustainable than one that tackles anticipated and specific impacts.

Accepting that both adaptation and poverty are complex, multi-dimensional and dynamic clearly implies there are no 'one-size-fits-all' solutions; that sustainable adaptation is a moving target, meaning different things in different contexts. Indeed Eriksen and O'Brien (2007) stress this. Sustainable adaptation then must also address how poverty itself is experienced and different forms of deprivation, and their social determinants and political contexts. In distinguishing between the chronic poor, transient poor and the non-poor, Tanner and Mitchell (2008) draw out some of the different types of adaptation measures which might be most appropriate, and these are shown in Table 5.3. For example, autonomous adaptation undertaken by the always poor (a sub-set of the chronically poor), such as selling assets, sending younger children to work or moving to more exposed areas, which could be classified as coping strategies, may make them more vulnerable and less able to cope with the next shock. Autonomous adaptation by the transient poor – such as diversifying livelihoods – may aid long-term adaptation and reduce poverty. This analysis, examining different types of poverty and different responses, can help to inform and qualify where, when, who, and what sustainable adaptation may be possible.

Tanner and Mitchell emphasise the threat of climate change to the world's poor. They recognise that to address chronic poverty and avoid rapid growth in numbers of chronically poor people, climate change adaptation that targets specific impacts must be accompanied by other institutional support mechanisms such as social protection, conflict prevention and delivery of services. They make a strong case for 'pro-poor adaptation' and that proactive adaptation has the potential to move people out of chronic poverty. But to achieve this, greater understanding of the inter-relationships between poverty, vulnerability and adaptation is needed. Only then can adaptation be effectively designed and implemented and supported to meet the needs of the poorest. In a similar vein, Shackleton et al. (2015) call for 'socially just adaptation' reviewing a series of case studies to demonstrate the multiple and interacting barriers to successful adaptation in sub-Saharan Africa. The next chapter looks specifically at how current understandings of movements in and out of poverty might be informed by

Table 5.3 Adaptation actions by poverty category

Aggregate category	Chronic poor		Transient poor		Non-poor
Specific poverty category	Always poor	Usually poor	Cyclical poor	Occasionally poor	Never poor
Autonomous adaptation	Selling of last assets; Sending younger children to work; Conflict, crime, sex work; Move to more exposed locations; Use fragile ecological assets	Intra-community transfer/charity; Sending children to work; Migration; Extended family; Conflict, crime, sex work	(Seasonal) migration; Less risky production; Working multiple jobs, longer hours; Investment in social capital/reciprocity	Diversify livelihoods; Investment in social capital/assets	Investments in multiple financial assets; Buy drought-tolerant seeds, new technologies; Diversify livelihoods; Invest in ethical/green options
Market-based adaptation		Promote micro-savings, micro-credit, micro-insurance; Cattle insurance	Weather-indexed insurance; Cattle insurance; Promote micro-savings, micro-credit, micro-insurance; Selling assets	Weather-indexed insurance; Promote micro-savings, micro-credit, micro-insurance; Selling assets	Crop insurance; Farm asset insurance and domestic insurances; Price hedging
Public policy-driven adaptation	Social pensions; Assisted migration; Democratisation of natural resource management/ecosystem rehabilitation; Promote health and nutrition services; Cash for work schemes; Conditional cash transfers	Community restocking; Subsidised seed banks; Democratisation of natural resource management/ecosystem rehabilitation; Promote health, disease reduction; Cash for work schemes	Community restocking; Improved remittances schemes; Subsidised seed banks; Improved climate information; Ecosystem rehabilitation; Irrigation schemes	Improved remittances schemes; Improved climate information; Employment assurance schemes; Social insurance schemes; Irrigation schemes and urban services provision	Taxation to pay for adaptation of the poor; Marketing of green agenda; Incentives for adaptation and mitigation choices

Source: Adapted from: Tanner & Mitchell 2008

resilience thinking, and charts more dynamic approaches to understanding poverty as noted in this work by Tanner and Mitchell (2008). Understanding adaptation as a more dynamic process – rather than an outcome of interventions – is also necessary to meet this challenge. An 'adaptation pathways' approach has recently been suggested to integrate these concerns.

Adaptation pathways

Maru and Stafford Smith (2014: 323) assert that 'it has taken 20 years to move the adaptation research agenda from a problem focus on impacts and vulnerabilities to a solutions-oriented focus on decision-making and risk management…'. They suggest re-framing adaptation as pathways, recognising that adaptation is a 'dynamic, long-term, transitory and transitional process that involves repeated decisions' (p322). In the same special issue of the journal *Global Environmental Change*, Wise et al. (2014) call for a paradigmatic shift in adaptation science and policy to conceptualise adaptation as an element of pathways of interacting global changes and societal responses. They detail the causes and consequences of current rationalist and impacts-oriented approaches to adaptation (see Box 5.2), to emphasise the need for this major change. As with the notion of sustainable adaptation, a pathways approach aims for a more dynamic, multiple-stressor understanding which recognises feedbacks between society and environment. They present pathways as a means of shifting focus from outcomes to processes of decision-making within a context of high uncertainty and inter-temporal complexity characterising a changing climate. In their understanding, a pathways approach enables the role and relations of incremental change to more systemic change to be opened up and negotiated. The adaptation pathways approach suggested by Wise et al. identifies five dimensions of adaptation underrepresented in contemporary adaptation approaches. The first, similar to calls for sustainable adaptation, asserts that climate change cannot be separated from other changes affecting society and environments. Second, it stresses the importance of cross-scale, cross-sectoral and cross-jurisdictional boundaries and accompanying threshold effects. Third, it asserts that feedbacks and inter-temporal dynamics must be recognised. Fourth, it contends that there are emergent properties in social ecological systems and great uncertainty in predicting trajectories of change. Finally, societal values, norms, rules and preferences have significant influence on the system.

Both the sustainable adaptation and adaptation pathways approaches highlight the spatial and temporal elements and potential of trade-offs in adaptation; in other words, action taken to address impacts experienced or even predicted now may have detrimental effects on other sections of society, in other locations or in the future. Adaptation actions undertaken now may close down future options; thus the adaptation pathways literature refers to the imperative of 'opening up' an 'adaptation space'.

Box 5.2 Implications and consequences of prevailing approaches to adaptation

Prevailing predict-provide and impact analytical approaches to adaptation have implications and consequences such as:

- Science is inappropriately used to try to resolve contested problem definitions and solutions.
- Solutions address symptoms and proximate causes – little scope for transformational change or to address root causes.
- Focuses on static measures of vulnerability and adaptation and fixed times; downplay temporal interdependencies.
- Limited opportunities for on-going learning.
- Emphasise management of specific quantifiable or observable risks through controlling environment – overlooks social, political and wider normative aspects of adaptation.
- Adaptation promoted as single or few decisions within a 'project' lifespan – not consideration of longer-term cultural, institutional, political and technological path-dependencies.
- Emphasis and investments in 'reducing uncertainty'.
- Lack of integration between climate and other drivers of change.
- 'Safe', tried and tested solutions work in favour of maintaining status quo, rather than innovation.
- Governments do not recognise own role and limitations – not participating as partners in learning and innovation, but as information, service and funding providers.
- Scientists seen as key producers of knowledge – learning framed by researchers rather than key adaptation stakeholders.

Source: Adapted from Wise et al. 2014

Jon Barnett and Saffron O'Neill refer to these detrimental impacts of climate change adaptation as maladaptation which they define as 'action taken ostensibly to avoid or reduce vulnerability to climate change that impacts adversely on, or increases the vulnerability of other systems, sectors or social groups' (Barnett & O'Neill 2010: 211). Their analysis of maladaptation identifies five pathways to maladaptation, where actions taken actually increase emissions of greenhouse gases; disproportionately burden the most vulnerable; have high opportunity costs; reduce incentives to adapt; or create path dependency. They also highlight that a critical problem dogging adaptation is the time lag between the changes in climate and the changes in institutions that often masks the longer-term and geographically and socially distant impacts of adaptation responses.

So, to an extent these latest writings on adaptation appear to be bringing together adaptation, vulnerability and resilience concepts to present a more dynamic and integrated approach, which attempts to frame adaptation within a broader context of sustainable development. The next sections look at how these ideas and concepts are understood and intertwined, and how a resilience perspective – or lens – might better inform adaptation to meet the challenges of sustainable development.

Distinguishing adaptation, vulnerability and resilience

What then does the climate change adaptation literature say about resilience? This section explores how adaptation, vulnerability and resilience are understood and applied, and how the literature distinguishes them. It reveals a number of different views of the relationships between these concepts. Janssen (2007) analysed these as separate sub-fields, or domains, in the global environmental change literature, but clearly there are linkages between them. Notably, many authors view resilience as the opposite, or antonym, of vulnerability and indeed this is a view often mirrored – as discussed in earlier chapters – in policy discourses. But they are distinct, if resilience is understood as a system property that provides an analytical lens through which to view and potentially assess or evaluate outcomes. I present examples of how resilience has been applied and how it provides insights and informs responses to climate change. Such an approach, I argue, leads to a focus on adaptive capacity as a concept around which it is possible to organise and mobilise adaptation.

Mark Pelling's work has been important here (see Pelling 2010). His approach, from human geography, has its roots in analysis of disasters and development often in urban settings. In the context of climate change adaptation, Pelling defines resilience as a refinement of actions to improve performance without changing guiding assumptions or the questioning of established routines. His framework for understanding adaptation distinguishes three visions of adaptation – representing three levels of engagement – which see adaptation as resilience, adaptation as transition, and adaptation as transformation. Thus Pelling views resilience as providing one approach to adaptation which is characterised by actions which seek to protect priority functions in the face of external threats. In this case, resilience approaches do not seek to realign development relations. Resilience becomes subsumed within adaptation in this interpretation, and implies a set of normative assumptions and applications. Here a resilience approach to adaptation is closer to 'adjustment adaptation' or perhaps 'reformist adaptation' described by Bassett and Fogelman (2013). Resilience here is about persisting and maintaining identity, about bouncing back.

In our 2007 paper (Nelson et al. 2007) we develop a quite different framework of how resilience is applied in much of the adaptation literature. In analysing how resilience ideas could inform adaptation we acknowledge these multiple meanings and the contested nature of the evolving theoretical and policy literature. But we argue that a resilience lens adds analytical depth and new dimensions to adaptation thinking and policy in a number of ways. In particular, and in line with the writings on sustainable adaptation or adaptation pathways, it recognises multiple threats or stressors – not just climate change and not specific impacts; it presents framework for integrating and acknowledging (if not fully understanding) dynamic and multi-scalar linkages; and, importantly, it holds the possibility of transformation. So in contrast to other writing, we do not see resilience as opposed to transformative change; a resilience lens can help to inform and identify the options for transformation. Thus Nelson et al. view resilience as a set of system characteristics – including the capacity to absorb change, capacity for learning, and self-organisation – which support or influence processes or different adaptations, including both incremental systems adjustments and transformation. In turn, these processes lead to a set of outcomes and different states of forms of *adaptedness*. This configuration is shown diagrammatically in Figure 5.1. Here, adaptedness is defined as 'a state in which a system is effective in relating with the environment and meets the normative goals of stakeholders' (Nelson et al. 2007: 400). Adaptedness is the outcome of adaptation processes, but it is never permanent, and changes in relation to context and in light of different disturbances. This understanding recognises adaptation as a process of deliberative change in anticipation of, or response to, external stimuli or stress. It might constitute either incremental adjustments, or transformative action. But core to the understanding is adaptation as the decision-making processes and the set of actions undertaken to maintain the capacity to deal with current or future change or shocks.

In this understanding, resilience does not lead to any one outcome; resilience is a set of system characteristics or capacities which will be applied in, or which will affect, different processes of change. Importantly, these are changing, including as a result of adaptation actions and their outcomes, and in response to different types of shocks or disturbances, as shown in Figure 5.1. In other words, there are feedback processes. Resilience is not

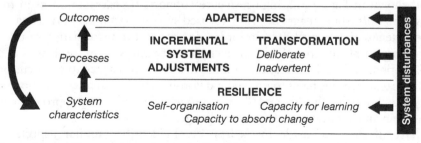

Figure 5.1 Resilience, adaptation and adaptedness

Source: Nelson et al. 2007

always a good thing or a healthy attribute; a highly resilient system may reside in undesirable states and may even be described as pathologically resistant to change. A system might be highly adapted to current conditions, which makes changes in response to disturbance difficult.

Reviewing the field of environmental change, Nelson et al. make a further distinction between adaptation and resilience, identifying separate streams of scholarship. Adaptation research is primarily actor-orientated in nature, focused on decision-making, on the roles of actors, their agency and actions. It is primarily prescriptive and normative and sees the overall objective of adaptation to maintain human well-being. In contrast, the resilience field is systems-orientated, and focused on analysing the relationships between different system components to understand the feedbacks and temporal dynamics. It is sometimes normative in its approach. But the overall objective is to maintain system capacity and flexibility in face of shocks and disturbance, but change is an inherent part of the system. The reconciliation of the actor-orientated and the systems perspectives remains a formidable challenge in the environmental change field, and it is interrogated in the exploration of resilience in this book.

Yet applying a resilience lens to understanding adaptation adds traction and insights in a number of ways (Nelson et al. 2007). It provides a dynamic understanding of change, particularly because it explicitly recognises feedbacks in the system. It enables a focus on identifying and developing sources of resilience in systems. Importantly it sees adaptation as not eliminating vulnerability – vulnerability exists in all systems, and the objective of adaptation is to understand the sources of vulnerability and identify acceptable levels of vulnerability and likely responses to different types of shocks or stressors. Therefore it differs from other conceptualisations that see resilience as the antonym of vulnerability, or adaptation as an end point.

Figure 5.2 shows how this framework is applied to understand resilience and adaptation – seeing resilience as a property of a system, and adaptation as a set of processes or responses, leads to insights into different actions and impacts.

In separating the resilience characteristics of the systems from the processes of adaptation, a more nuanced understanding of the relationship and the role of resilience emerges. Hence in the examples used in Nelson et al.'s analysis and shown in Figure 5.2, high resilience supports transformation as in the case of a shift from an agriculturally-focused system to tourism in Arizona. In the system with low resilience, in Jordan, inadvertent transformation resulted in agricultural collapse. This distinction, between system resilience and adaptation, and analysis is important because the next two chapters focus on how a resilience lens helps to understand non-incremental change and moving beyond 'bouncing back', and how a re-visioned resilience approach can underpin and support transformative change.

Overall the case made by Nelson et al. identifies similar aspects of resilience thinking as highlighted earlier. They point to four areas in particular where a resilience framework can contribute to adaptation studies.

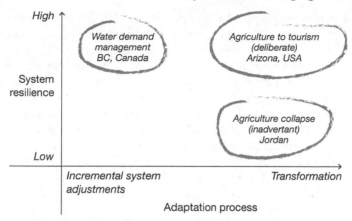

Figure 5.2 Relationships between system resilience and adaptation processes

Source: Nelson et al. 2007

The first is its recognition of multiple states, which brings to adaptation approaches a consideration of options – as the discussion of adaptation pathways emphasised, the need to open up rather than close down future options – and a move away from a 'desirable' equilibrium. It also raises important questions about what constitutes a 'desired state' and how this can be identified and negotiated between different interests and societal actors. The issue of thresholds is also highlighted, when the possibility of multiple states is acknowledged, again emphasising the non-linear nature of change and the possibility of abrupt change. Second, the framework puts emphasis on adaptive capacity as a pre-requisite for adaptation. It asks the question, what is needed for successful adaptation? Adaptive capacity relates to resources, but if we understand adaptive capacity as latent characteristics of a system then this prompts consideration of what might trigger adaptation, where the opportunities and surprises are. Third, a resilience framework draws attention to the possible trade-offs between resilience and adaptiveness. Whilst conventional adaptation approaches emphasise adjustments to move towards a desire state that reduces risks, resilience puts greater attention on building capacity to cope with future change. A system highly adapted to current situation may have weakened capacity to cope with change in circumstances. Finally a resilience approach can inform governance and normative issues, especially by drawing attention to the role of adaptive management and governance, the definition and negotiation of a desired state and issues such as equity, diversity and justice.

Applying the resilience lens

In Chapter 4 the findings of my research in Tanzania and Mozambique with Matthew Bunce and Sergio Rosendo (2010) highlighted how multiple stressors interacted to undermine peoples' livelihood security. It showed how policies enacted by governments and donors designed to support sustainable development actually had regressive impacts, particularly on the poor, in the coastal communities we studied. We argued that the livelihood resilience of these people was undermined not only by a series of interacting stressors – many of which are completely beyond their control and beyond even the power of their governments – but also by the very policies implemented with an explicit aim of 'development'. The discussion in Chapter 4 centred on the implications of the findings for understanding the social dynamics of resilience – including the 'winners and losers' and the cross-scale dimensions of resilience and vulnerability. We have posited a number of reasons for this 'policy misfit' (Bunce et al. 2010). These include, first, that development and conservation policies do not take long-term climate change into account – climate change is considered separately. In fact in many instances, development and conservation agencies are locked into short-term planning and 'project' cycles, so do not, and do not have capacity and resources to, look at longer-term changes and drivers. Secondly, these policies often reflect international priorities and funding streams, rather than local priorities for livelihoods. Third, they do not take account of interactions between multiple stressors, in other words the dynamic context in which they are implemented, and they address one issue only, for example, biodiversity conservation or water scarcity. Finally, they take a very static view of change, extrapolating from the present and making assumptions about the linear trajectory of change.

Insights from resilience might overcome some of the weakness in current adaptation and development, particularly to understand adaptation and its outcomes and the reasons for maladaptation better. Thus I argue that a socially-informed resilience lens might contribute to formulating more sustainable adaptation actions and policy. To develop a similar approach we undertook an analysis of nine climate change responses and assessed their impact on the resilience of the social ecological systems within which they were embedded (see Adger et al. 2011). From this analysis we were able to identify three critical factors that influence how the responses affect short-term coping versus long-term sustainability. These are first, the framing of the problem to be addressed; second, the type of governance structures that implement the responses; and third, the extent to which feedbacks are recognised and incorporated into the action.

We demonstrated that how the issue to be addressed is framed and conceptualised affects the impacts of the response or adaptation measure on system resilience. So, narrowly defined, technologically-focused framing of the problem – often in terms of a single risk, climate change impact and single location or sector – considers a limited number of possible responses to identified threats. Conversely, broader and more inclusive frames recognise the importance of other system drivers and the maintenance of response flexibility, and tend to focus on the management of an integrated set of problems rather than on the implementation of specific actions. From the nine examples of adaptation to climate change we assessed, the responses to pine bark beetle infestation and declining fish stocks are examples of narrow, technological approaches. In western Canada, activities are designed to maximise short-term economic output of forest resources. In Uganda government action is directed at controlling fishing technologies. In contrast, water resources management in Brazil addresses the multiple development and economic drivers of water use, moving beyond a supply and demand approach. In the Cayman Islands responses to hurricanes focus on tropical storm preparedness, but also incorporate an active learning cycle and wider governance issues to build wider system resilience which enhances capacity to respond to a wide set of – as yet unknown – disturbances.

Governance arrangements were also found to influence the resilience of climate change responses. Formal governance arrangements include vertical, top-down approaches, as well as decentralised and more horizontally integrated systems. Thus drought management in Northeast Brazil, a highly vertical approach to governance has been detrimental to the emergence of more local and expansive institutions that can represent adaptive capacity. These top-down actions are often associated with narrow technologically defined problems. Often more broadly conceived and framed adaptations, such as coastal management in the UK, are open to participation and contributions from a wider set of social actors and stakeholders at different scales. These actors may even be involved in problem identification and formulating responses. This is often not the case, and as in the examples from Tanzania and Mozambique discussed in Chapter 4, the cross-scale dimensions and the differing access to decision-making and power asymmetries militate against such inclusion. The differences in temporal and spatial scales, between where decisions are made and what and who is affected, is illustrated by the case of biofuels promotion, which is linked with increased deforestation in the Amazon and changes in land cover and soil conservation management in the USA.

In this way, the effects of each response may be displaced over time or space. This in turn will affect whether and how feedbacks are taken account of and reacted to. Social learning is more likely if feedbacks occur soon relative to action and if those most affected by feedbacks are those responsible for the action. Slow feedbacks, those that are spatially distant, or those that are masked, or ignored, by short-term economic, political or productivity gains, are less likely to affect policy responses. In the biofuels example, ecological impacts are masked by short-term economic gain and are dislocated through space. Drought responses in Brazil and Canada mask ecological impacts through government programmes and transfers.

Understanding adaptive capacity

In Brown and Westaway (2011) we posited the concept of adaptive capacity as a meeting point between social ecological systems and human development knowledge domains, with related concepts of resilience, well-being, capacity, and capabilities potentially able to inform a dynamic and nuanced view of agency in environmental change. Similarly Nathan Engle (2011: 655) suggests adaptive capacity is a 'common thread between vulnerability and resilience frameworks' that has powerful traction for policy and governance of adaptation to climate change. In making the case for assessments of adaptive capacity which include a more dynamic systems perspective, he recognises the difficulties in gauging adaptive capacity. First, because adaptive capacity may be a latent property which only comes into play and is evident in response to a specific event. Thus many studies of adaptive capacity are historical, investigating past events. Second, there are no standard measures, often assessment uses a range of indicators to formulate indices, making adaptive capacity difficult to operationalise. I examine here some of the approaches to assessment and specifically what these mean for adaptation actions and development. Engle notes that many studies of adaptive capacity are descriptive, and that few analyses reveal details on dynamics and determinants of adaptive capacity. Eakin et al. (2014) suggest that while research has made considerable progress in identifying the attributes of a population indicative of overall capacity (much in line with first wave resilience work discussed in Chapter 3), there is much less certainty on how these attributes work together. They suggest that some attributes need more attention, especially if climate change adaptation is to work synergistically with sustainable development.

Adaptive capacity consists of the pre-conditions necessary to enable adaptation to take place, where *adaptation* is a process or activity undertaken in order to alleviate the adverse impacts of environmental stresses or take advantages of new opportunities. Adaptive capacity is thus a *latent characteristic* which must be activated in order to effect adaptation; Nelson et al. (2007) define adaptive capacity as the set of resources, and the ability to employ those resources, that are pre-requisites to adaptation.

McClanahan et al. (2009) undertook a synthesis of adaptation literature, reviewing the various attributes and factors that had been identified as contributing towards adaptive capacity. Six key factors emerged as important pre-requisites for successful adaptation: recognition of the need to adapt; belief that adaptation is possible and desirable; willingness to undertake adaptation; availability of resources necessary for implementation of adaptation measures; ability to deploy resources in an appropriate way; and external enablers to implementation. Five of these attributes describe characteristics of individuals and households, only one describes characteristics of external context. These factors are similar to those detailed by Dolan and Walker (2004) who identify six components of community scale adaptive capacity: income and its distribution; access to technology; information and skills available; risk perception and awareness; social capital; and institutional frameworks (including things like insurance, property rights, planning processes).

Many assessments focus on assets or capitals at different scales. For example, Nelson at al. (2010) adapt the five capitals of the Sustainable Livelihoods framework to measure adaptive capacity based on farm household surveys. They compile the following indicators to assess adaptive capacity:

- human capital – education enrolment and attainment
- social capital – community groups and internet connections
- natural capital – land use and land use change, protected areas, etc.
- physical capital – infrastructure, transport
- financial capital – per capita incomes and regional incomes.

Jones et al.'s work (2010) for the Overseas Development Institute (ODI) working with a number of development NGOs and the African Climate Change Resilience Alliance, presents a Local Adaptive Capacity framework that aims to move beyond the list of attributes to develop a more local systems perspective on adaptive capacity. They identify five characteristics of adaptive capacity, shown in Table 5.4, but they stress that this is not a framework for describing an adaptive system, but one for understanding what features support adaptive capacity and, importantly, the relationships between these features. Based on local level consultations, the characteristics influence and determine the degree to which a community is resilient and responsive in the face of changes in its external environment. The overall aim of this analysis is to work with development initiatives and make sure that interventions to support adaptation to climate change are synergistic with development and will ultimately address underlying drivers of vulnerability and poverty.

Table 5.4 Local adaptive capacity characteristics

Characteristic	Features that reflect a high adaptive capacity
Asset base	Availability of key assets that allow the system to respond to evolving circumstances
Institutions and entitlements	Existence of an appropriate and evolving institutional environment that allows fair access and entitlement to key assets and capitals
Knowledge and information	The system has the ability to collect, analyse and disseminate knowledge and information in support of adaptation activities
Innovation	The system creates an enabling environment to foster innovation, experimentation and the ability to explore niche solutions in order to take advantage of new opportunities
Flexible forward-looking decision-making and governance	The system is able to anticipate, incorporate and respond to changes with regards to its governance structures and future planning

Source: Jones, Ludi & Levine 2010: 4

Assessment of adaptive capacity and its characteristics in a given context can provide a more nuanced view of what policies might be most appropriate, and which might support both adaptation and development. The analysis we undertook in the western Indian Ocean attempted to understand how different approaches to conservation and development might be affected by, and in turn affect, adaptive capacity (McClanahan et al. 2008; Cinner et al. 2012). We undertook empirical analysis of adaptive capacity based on data from more than 1,500 households in 42 communities in five countries. We developed an index of adaptive capacity based on eight indicators, which were weighted according to expert poll. The indicators and the characteristics of adaptive capacity are shown in Table 5.5.

We undertook an analysis of environmental susceptibility to coral bleaching, based on characterisation of reef systems throughout the region, and compared this with the adaptive capacity of communities who depended on these reefs for fisheries. Environmental susceptibility was a compound indicator of how likely reefs were to be damaged as a result of a bleaching event. We postulated that these two characteristics of the marine social ecological system – the environmental susceptibility of the reef, and the adaptive capacity of fishing communities reliant on it – would be important in shaping and identifying effective conservation and adaptation policies.

The general approach and classification we developed in shown in Figure 5.3a. Thus conventional conservation approaches –'protect and preserve' – are only likely to be effective in sites where there is high adaptive capacity, for example, where people have access to alternative livelihoods, and there

Table 5.5 Elements of an index of adaptive capacity

Indicator	Measurement
Recognition of causality and human agency in marine resources	If interviewee suggested factors that affect fish populations and/or interventions to improve populations
Capacity to anticipate change and develop response strategies	Stated response of fishers to hypothetical 50% decline in catch
Occupational mobility	Changes in employment in last 5 years; whether forced or voluntary and if new occupation preferred
Wealth	Presence of 15 material assets
Occupational multiplicity	Total number of person-occupations per household
Social capital	Whether interviewee member of community organisation
Technology	Number of different fishing gears used
Infrastructure	Presence of 20 different infrastructure items in community

Source: Adapted from McClanahan et al. 2008

is high social capital and infrastructure support such as local government services, schools and health facilities. For sites where adaptive capacity is low, conservation is unlikely to work and will have socially deleterious effects; in fact it is likely to be both socially unjust and unsupported. Depending on the environmental susceptibility, either relief and social protection, or intense capacity building efforts are necessary. Adaptation and transformation are more appropriate and likely to be most effective in areas where both adaptive capacity and environmental susceptibility are high (the upper right quadrant in Figure 5.3a).

When these sites were mapped empirically onto this diagram we could then see where and how current policies might or might not be appropriate, for both conservation planning and longer-term adaptation (Figure 5.3b). It

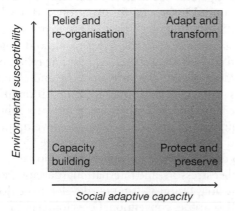

Figure 5.3a Conceptual model of gradients of adaptive capacity and environmental susceptability and suggested governance strategies

Source: McClanahan et al. 2008: 54

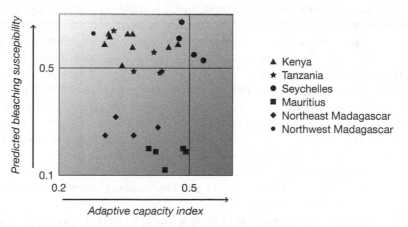

Figure 5.3b Empirical case study from five countries in the western Indian Ocean

Source: McClanahan et al. 2008: 54

became apparent where particular countries were clustered on the grid, and how the current approaches to conservation and adaptation compared with what might be considered most appropriate for the specific conditions.

We later further developed the analysis to extend the consideration of ecological characteristics of coral reef systems (McClanahan et al. 2009), and vulnerability of reef-based communities to change (Cinner et al. 2012). We were then able to develop a much more detailed analysis of possible policy options – in short-, medium- and long-term, to address the different vulnerability contexts in the region (see Table 2 in Cinner 2012).

What this analysis reveals is that many policies that aim to build adaptive capacity correspond to conventional development actions, investing in health, local governance and poverty reduction in the longer term, strengthening local community groups and social networks, and improving infrastructure in the medium term; and improved information and market information in the shorter term. Some measures to address exposure are closely aligned with relief and risk reduction in the short-term, and might include emergency evacuation and conventional relief, e.g. famine relief, but in the long term would invest in developing alternative energy and climate mitigation. This analysis 'opens up' options for adaptation, ultimately linking to mitigation of climate change, and gives insights into where adaptation to climate change might correspond to broader development strategies, or where more targeted responses might be appropriate.

These types of analyses start to decipher the complex relationship between adaptation to climate change and other stressors, and international development, moving beyond the approach that adaptation is 'simply' development done well. Currently, as Eakin et al. (2014) argue, conventional development policy and climate change adaptation target different capacities. They distinguish a set of 'generic' capacities and 'specific' capacities that affect adaptation (shown in Table 5.6). Generic capacities are typically the focus of development policy, whereas adaptation targets specific capacities. This helps to explain why climate change adaptation is not effectively addressing poverty and why development does not equate to adaptation. But there is a crucial need to understand the interactions and relations between these in order for interventions to build adaptive capacity and resilience.

Eakin et al. suggest a heuristic to analyse the interplay between generic and specific capacities for a specific social group (Figure 5.4). They propose that disaggregating capacity into these specific and generic attributes and exploring their relation across time, space and scale can support design and implementation of adaptation policy, especially in conditions of high uncertainty (Eakin et al. 2014). This has some similarities to the grid we developed for the analysis of adaptive capacity and reef communities in the western Indian Ocean (Figure 5.3b).

In Figure 5.4, the current state of the social system is shaped by broader societal institutional arrangements (system-level capacities) which address human development needs (generic capacities) and risk (specific capacity).

Table 5.6 Generic capacities for development and specific capacities for climate change adaptation

	Individual actor	*System-level*
Generic	Income level and structure savings Material assets Health status Education levels Population mobility Participation in social organisations	Economic productivity Information infrastructure Poverty levels Economic and social inequality Transparency in governance Population-level education Sanitation Health care services Built environment integrity
Specific	Climatic information use Protection of private property Climate risk insurance Adoption of technologies (e.g. crop varieties) to reduce climate impacts Cultural climate prediction Traditional risk mitigation strategies	Insurance provisioning systems Early warning systems Scenario development Infrastructure investment Disaster planning Disaster compensation funds Risk mitigation planning

Source: Eakin et al. 2014: 2

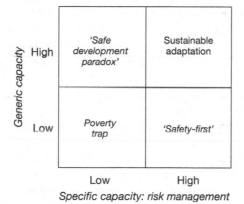

Figure 5.4 Capacities matrix

Source: Eakin et al. 2014: 4

When both generic and specific capacities are low (lower left quadrant in Figure 5.4) society may be in a poverty trap, with chronic and intense stress, and positive feedbacks between capacities, making escape from the trap and constructive change difficult. The next chapter explores the resilience insights into these traps. In the 'safety-first' quadrant, societies might have developed high capacity to deal with specific risks, but have low generic capacity. Households here tend to be asset poor, but have developed coping capacities and are often able to smooth their consumption needs in highly variable environments. Their circumstances may lead them to prioritise short-term risk minimisation over longer-term sustainability and welfare – they conform to classic 'risk-minimising peasants' (Ellis 1988), or erroneously, 'the resilient poor'. The 'safe development paradox' domain (upper left) has populations with higher generic capacities – so higher conventional development indicators such as assets or education – but are poorly equipped to deal with particular risks. Eakin et al. (2014) cite the example of Hurricane Katrina in Louisiana, USA, where federal policies designed to address hazards and make development safe actually undermined local capacities for effective risk management. The next chapter discusses this in terms of a rigidity trap. Only in the context of high generic and high specific capacities, and where development and adaptation policies are reinforcing, is sustainable adaptation and perhaps transformation a possibility.

Much of the literature points to the importance of socio-cognitive, cultural and other factors, as well as agency, and these are not captured in most of the existing measures of adaptive capacity. Additionally, transformative change might require a different set of capacities and that some factors – such as place attachment and cultural or occupational identity – may support adaptation but act as a barrier to transformation. Marshall et al. (2012) explore the tensions and possible trade-offs between capacities for adaptation and transformation. We show how different characteristics of adaptive capacity played out when societies were faced with the option of transformative change. We found that factors that strengthen adaptive capacity – for example, place attachment and aspects of identity – might actually work against transformative change. The empirical example we used was peanut farmers in Australia, faced with impacts of climate change that made the continuation of their agricultural practices unviable. Would they move locations to more favourable environment to continue farming peanuts? Attachment to place was a critical factor, and the implications of this are explored further in the final chapter.

In summary, there are significantly different understandings of the relationship between adaptation, resilience and transformation and this shapes both research and policy and practice. Some authors see a focus on resilience as aligned with adaptation approaches designed to reduce vulnerability and strengthen the ability to withstand and absorb shocks, but essentially to stay the same. This is close indeed to the policy applications of resilience identified in Chapter 2. However, a view of resilience as a system property which affects the ability to deal with multiple stressors and uncertainty, and which recognises multiple feedbacks, emphasising learning, self-organisation and adaptive capacity, provides a different way of understanding the relationship between adaptation and resilience. As an analytical lens, resilience does not prescribe a certain type of adaptation. But in emphasising capacities, it can also be fundamental to our understanding of non-incremental change and transformation. This also brings the differential capacities of different people into focus, and highlights the role of interventions in building or undermining capacities. The following chapters look at how resilience concepts inform understanding of poverty and specifically poverty eradication; and at transformational change and how a re-visioned resilience can support transformative change necessary to address chronic poverty and tackle climate change.

References

Adger WN, Huq S, Brown K, Conway D and Hulme M (2003) Adaptation to climate change in the developing world. *Progress in Development Studies*, 3(3), 179–95.

Adger WN, Brown K, Nelson DR, Berkes F, Eakin H, Folke C and Tompkins EL (2011) Resilience implications of policy responses to climate change. *Wiley Interdisciplinary Reviews: Climate Change*, 2(5), 757–66.

Adger WN, Barnett J, Brown K, Marshall N and O'Brien K (2012) Cultural dimensions of climate change impacts and adaptation. *Nature Climate Change*, 3(2), 112–17.

Ayers J (2010) Understanding the adaptation paradox: Can global climate change adaptation policy be locally inclusive? PhD thesis. London School of Economics and Political Science.

Barnett J and O'Neill S (2010) Maladaptation. *Global Environmental Change*, 20(2), 211–13.

Bassett T and Fogelman C (2013) Déjà vu or something new? The adaptation concept in the climate change literature. *Geoforum*, 48, 42–53.

Brown K (2011) Sustainable adaptation: An oxymoron? *Climate and Development*, 3(1), 21–31.

Brown K and Westaway E (2011) Agency, capacity, and resilience to environmental change: Lessons from human development, wellbeing, and disasters. *Annual Review of Environment and Resources*, 36(1), 321–42.

Bunce M, Brown K and Rosendo S (2010) Policy misfits, climate change and cross-scale vulnerability in coastal Africa: How development pojects undermine resilience. *Environmental Science and Policy*, 13(6), 485–97.

Burton I, Kates R and White G (1978) *The environment as hazard*. New York: Oxford University Press.

Cinner J, McClanahan T, Daw T, Mainad J, Steadf SM, Wamukotad A, Brown K and Bodinh O (2012) Vulnerability of coastal communities to key impacts of climate change on coral reef fisheries. *Global Environmental Change*, 22(1), 12–20.

Dolan A and Walker I (2004) Understanding vulnerability of coastal communities to climate change related risks. *Journal of Coastal Research*, SI 39 (Proceedings of the 8th International Coastal Symposium), 1316–23.

Eakin H, Lemos M and Nelson D (2014) Differentiating capacities as a means to sustainable climate change adaptation. *Global Environmental Change*, 27, 1–8.

Ellis F (1988) *Peasant economics: Farm households in agrarian development*. Cambridge: Cambridge University Press.

Engle N (2011) Adaptive capacity and its assessment. *Global Environmental Change*, 21(2), 647–56.

Eriksen S and Brown K (2011) Sustainable adaptation to climate change. *Climate and Development*, 3(1), 3–6.

Eriksen S and O'Brien K (2007) Vulnerability, poverty and the need for sustainable adaptation measures. *Climate Policy*, 7(4), 337–52.

Eriksen S, Aldunce P, Bahinipati CS, Martins RDA, Molefe JI, Nhemachena C, O'Brien K, Olorunfemi F, Park J, Synga L and Ulsrud K (2011) When not every response to climate change is a good one: Identifying principles for sustainable adaptation. *Climate and Development*, 3(1), 7–20.

Fransen TS, Nakhooda S, Kuramochi T, Caravani A, Prizzon A, Shimizu N, Tilley H, Halimanjaya A and Welham B (2013) *Mobilisation International Climate Finance: Lessons from the Fast-start Finance Period*. Washington, DC: Global Environmental Strategies, and Open Climate Network.

Janssen M (2007) An update on the scholarly networks on resilience, vulnerability, and adaptation within the human dimensions of global environmental change. *Ecology and Society*, 12(2), 9.

Jones L, Ludi E and Levine S (2010) Towards a characterisation of adaptive capacity: A framework for analysing adaptive capacity at the local level. *ODI Background Notes*, (December), 1–7.

McCarthy, James J (2001) *Climate change 2001: Impacts, adaptation, and vulnerability: contribution of Working Group II to the third assessment report of the Intergovernmental Panel on Climate Change*. Cambridge: Cambridge University Press.

McClanahan T, Cinner J, Maina J, Graham NAJ, Daw TM, Stead SM, Wamukota A, Brown K, Ateweberhan M, Venus V and Polunin NVC (2008) Conservation action in a changing climate. *Conservation Letters*, 1(2), 53–9.

McClanahan T, Cinner J, Graham NA, Daw TM, Maina J, Stead SM, Wamukota A and Brown K (2009) Identifying reefs of hope and hopeful actions: Contextualising environmental, ecological, and social parameters to respond effectively to climate change. *Conservation Biology*, 23(3), 662–71.

Marshall N, Park S, Adger WN, Brown K and Howden M (2012) Transformational capacity and the influence of place and identity. *Environmental Research Letters*, 7(3), 034022, 9.

Maru Y and Stafford Smith M (2014) GEC special edition: Reframing adaptation pathways. *Global Environmental Change*, 28, 322–24.

Mearns R and Norton A (2010) *Social dimensions of climate change: Equity and vulnerability in a warming world*. Washington DC: World Bank Publications.

Nelson D, Adger WN and Brown K (2007) Adaptation to environmental change: Contributions of a resilience framework. *Annual Review of Environment and Resources*, 32(1), 395–419.

Nelson R, Kokic P, Crimp S, Martin P, Meinke H, Howden S, De Voil P and Nidumolu U (2010) The vulnerability of Australian rural communities to climate variability and change: Part II—integrating impacts with adaptive capacity. *Environmental Science and Policy*, 13(1), 18–27.

Pelling M (2010) *Adaptation to climate change: From resilience to transformation*. Abingdon and New York: Routledge.

Shackleton S, Ziervogel G, Sallu S, Gill T and Tschakert P (2015) Why is socially-just climate change adaptation in sub-Saharan Africa so challenging? A review of barriers identified from empirical cases. *Wiley Interdisciplinary Reviews: Climate Change*, 6(3), 321–44.

Tanner T and Mitchell T (2008) Entrenchment or enhancement: Could climate change adaptation help to reduce chronic poverty? *IDS Bulletin*, 39(4), 6–15.

UNFCCC (n.d.) Adaptation Fund. Available from: http://unfccc.int/cooperation_ and_support/financial_mechanism/adaptation_fund/items/3659.php (accessed 3 October 2015).

UNFCCC (n.d.) Green Climate Fund. Available from: http://unfccc.int/cooperation_ and_support/financial_mechanism/green_climate_fund/items/5869.php (accessed 3 October 2015).

UNFCCC (n.d.) Least Developed Countries (LDC) Fund. Available from: http:// unfccc.int/cooperation_and_support/financial_mechanism/least_developed_ country_fund/items/4723.php (accessed 3 October 2015).

UNFCCC (n.d.) Nairobi work programme on impacts, vulnerability and adaptation to climate change (NWP). Available from: https://www3.unfccc.int/pls/apex/ f?p=333:1:213629852078229 (accessed 3 October 2015).

UNFCCC (n.d.) National Adaptation Programmes of Action (NAPAs). Available from: http://unfccc.int/national_reports/napa/items/2719.php (accessed 3 October 2015).

Watts M (1983) On the poverty of theory: Natural hazards research in context. In: Hewitt K (ed.), *Interpretation of calamity: From the viewpoint of human ecology.* Boston, MA: Allen & Unwin, 231–62.

Wise R, Fazey I, Smith M, Park SE, Eakine HC, Archer Van Garderen ERM and Campbell B (2014) Reconceptualising adaptation to climate change as part of pathways of change and response. *Global Environmental Change*, 28, 325–36.

6 Traps and transformations
The resilience of poverty

What can resilience ideas contribute towards understanding poverty and its persistence and elimination, and how helpful is resilience in thinking about international development in the face of global change? In earlier chapters, I show how resilience has become an important part of the lexicon of international development policy (especially with regard to climate change and disasters), but does resilience have greater analytical depth and applicability to inform development more widely – both academically and in terms of policy and practice? As noted, resilience concepts have been widely adopted in disaster relief and humanitarian aid (e.g. DFID 2011, Ramalingam 2013) and in climate change adaptation and international development. In Chapter 5 I discuss how resilience might shed light on the processes of adaptation and how it adds to analysis of climate change adaptation. In drawing attention to adaptive and transformative capacities, it might help to overcome some of the potential pitfalls for current approaches to climate change adaptation. In both the policy world and in scientific literature, resilience is seen as a boundary concept and a bridge linking development and climate change agendas, but could it provide more significant insights than this?

This chapter extends the analysis of how resilience – in addition to providing insights into the interactions of multiple stressors, and more systemic thinking about sustainability – might provide some insights into understanding development, and suggest how poor people can build and sustain secure livelihoods in times of rapid change. It addresses two questions: first, what does resilience tell us about poverty and particularly poverty traps? In particular, how does it contribute to thinking about poverty, and its persistence and elimination in a more dynamic and relational way within the context of environmental change? Second, how does resilience help us to understand processes of change and particularly transformative change? Such changes might involve changes to institutions, cultures and agency, empowering people to envisage and implement alternative pathways. How can different perspectives on transformation help to inform development in the era of global change? Finally, I suggest what the implications of this shift in thinking might be for conventional development policy. I start by looking at how recent analyses have integrated resilience concepts into understanding poverty and poverty alleviation, and what resilience science adds to poverty analysis.

Understanding poverty dynamics

Are the poor 'resilient'? Chapter 4 discusses how the lived experiences of resilience reveal some of social and political dynamics and the differentiated nature of how people deal with change, and how they articulate their experiences and their capacities, and the effects of externally driven (well-intentioned) policies and interventions on their vulnerabilities. People are both resilient and vulnerable, not simply one or the other, and the everyday forms of resilience displayed highlight the relational and transactional nature of capacities, contingent on agency, power and access. These factors are commonly underplayed in conventional resilience analysis but they are central to understanding social and economic development. Acknowledging these aspects highlights that there are costs – especially social costs – to resilience, and that resilience is socially contingent, but also offers suggestions of how resilience analysis might be extended to address issues of poverty and deprivation. There is still a strong discourse around the resilience of the poor, and this chapter explores some of the different dimensions of this. To an extent it is in one sense the persistence of poverty itself which poses challenges to international development, and efforts to eliminate or alleviate deprivation and poverty. I use this chapter to develop in more depth ideas of how resilience concepts can inform international development in the face of changes – both gradual and slow, and sudden and unpredictable. These changes include climate change and other environmental changes – biodiversity loss, large-scale land conversion, urbanisation, water depletion – and also other social and economic changes associated with migration, globalisation, changing markets and technology, as well as financial shocks.

A number of recent analyses apply a resilience approach to understanding relationships between poverty (or poor people), natural resource management and change. But my intention here is to review resilience concepts and analyses which give traction and add to existing knowledge of poverty. So I make a distinction between undertaking resilience-focused studies of poor people and their management of natural resources or their adaptation action, and a more nuanced fusion of poverty and resilience concepts. I use resilience to add traction to understanding poverty, its persistence and its elimination.

Poverty is generally understood as 'the lack of, or inability to achieve, a socially acceptable standard of living, or the possession of insufficient resources to meet basic needs' (ESPA 2012).[1] Of course, the definition of a socially acceptable standard of living and even basic needs are socially contingent, and may be understood and constructed differently according to context. A distinction is conventionally made between absolute and relative poverty, where absolute poverty is the inability to meet absolute minimum requirements for human survival, and the poverty of an individual or a household is considered independently from others. Poverty lines or levels are delineated by governments and multilateral agencies, and people in poverty are defined as those falling below this. The poverty line often represents a minimum level of household income; for example, the most well-known is the global comparative figure of $1.25 a day (Ravallion et al. 2009). But it might also be defined as lack of other basic needs, such as access to safe water, minimum health care, malnutrition or food security. Relative poverty considers the status of an individual or household in relation to others within a given context. This brings a focus on distributional aspects of poverty. Many poverty analyses describe the *condition* of being poor, rather than how or why the condition exists.

In the past two decades there have been significant advances in how poverty is understood and analysed and this has had significant impacts on international policy and practice. Addison et al. (2009) provide an overview of 'three main fronts on which progress must be made if we are to deepen our understanding of why poverty occurs and significantly improve the effectiveness of poverty reduction policies' (p3). They claim that poverty research must focus on poverty dynamics, analysing changes over life course and across generations. Second, poverty must be understood as multi-dimensional in nature, moving beyond a focus primarily on income to encompass objective, subjective and relational dimensions of well-being. Third, these advances necessitate increasing work across and between a wider range of disciplines, moving away from an economics dominated understanding of poverty. Their edited volume emphasises the dynamics of poverty as the critical frontier for development thinking and practice. They claim that only in the last decade has poverty research seriously and systematically looked at time and dynamics, and that until the mid-1980s the main ways in which poverty research accounted for time was by looking at trends in income and economic indicators, seasonality and limited historic accounts of poverty.

Literature on poverty classically distinguishes between transitory and chronic poverty, as shown in Table 5.3 in Chapter 5 related to climate change adaptation measures. Chronic poverty is poverty of extended duration, which may last a lifetime and be transmitted from one generation to another (Bird & Higgins 2011). It has been a major focus of research – for example through the Chronic Poverty Research Centre[2] – and aid donor investments and programmes. Green (2009) presents an anthropological perspective on chronic poverty arguing that durable poverty and destitution highlight the institutional nexus of social relations that ensure that certain people are likely to experience poverty for an extended period. She makes the case that reconsidering chronic poverty as destitution recognises the depth and duration of poverty as an outcome of social exclusion and as social rather than economic status. Green stresses the resilience of poverty itself, its tenacity brought about by social institutions which perpetuate relations of exclusion and allocation, ensuring that certain individuals and households remain poor. Furthermore, the empirical evidence suggests that the longer you are poor, the more likely you are to remain so. The durable poor are often the most severely poor, and are caught in trans-generational poverty traps which deepen vertically and horizontally across time and space.

Krishna's analysis (2009) uses participatory methods and subjective assessments to analyse poverty dynamics, bringing together studies from more than 200 communities in seven different sites in five countries in a 'stages of progress method'. It looks at trends in poverty dynamics over a 25-year period, identifying households moving in and out of poverty, using a participatory approach to identify poverty thresholds (a similar approach taken by Béné et al. 2011). Significant proportions of households have escaped poverty in the last 25 years (up to 24 per cent, see Table 8.2 p189), but during the same period large numbers of households have fallen into poverty (up to 19 per cent). Krishna (2009) looks at some of the reasons for these movements in and out of poverty; in general, multiple linked factors propel households into poverty, although health and health-related expenses is the single most important reason in each region (see Table 6.1). Diversification of income is shown to be the principle means of escaping poverty (see Table 6.2), building on a major body of literature around livelihood diversification (Ellis 2000).

Table 6.1 Principle reasons for falling into poverty (per cent of descending households)

Reason	Rajasthan, India n = 364	Gujarat, India n = 189	Western Kenya n = 172	Andhra, India n = 335	Uganda: central and western n = 202	Peru: Puno and Cacamarca n = 252
Poor health and health-related expenses	60	88	74	74	71	67
Marriage/dowry/ new household-related expenses	31	68		69	18	29
Funeral-related expenses	34	49	64	28	15	11
High-interest private debt	72	52	60			
Drought/irrigation failure/crop disease	18			44	19	11
Unproductive land/ land exhaustion		38			8	

Source: Krishna 2009: 190

Table 6.2 Principle reasons for escaping poverty (per cent of escaping households)

Reason	Rajasthan, India n = 499	Gujarat, India n = 285	Western Kenya n = 172	Andhra, India n = 348	Uganda: Central and Western n = 398	Peru: Puno and Cacamarca n = 324
Diversification of income	70	35	78	51	54	69
Private sector employment	7	32	61	7	9	19
Public sector employment	11	39	13	11	6	10
Government assistance/ NGO scheme	8	6		7		4
Irrigation	27	29		25		

Source: Krishna 2009: 192

Davis' analysis (2009) uses a life history approach to explain this multi-causality in more detail, examining the trajectory patterns of households in Bangladesh. It identifies key 'last straw' threshold effects, other cumulative effects and how poverty outcomes are critically based on the order or sequence of events and complex interactions between them. Davis observes (rather differently to Krishna) that improvements in peoples' life conditions tend to happen only gradually, whereas sudden declines are more common. Crises are more likely to produce a serious decline when they directly affect a constitutive aspect of peoples' well-being – such as health – or when someone has few buffers (as a very poor person). Again this identifies illness as a key factor, but other important factors are found to be the death of a productive household member and accidents. These studies empirically demonstrate some core resilience principles, including linkages and multi-causality and dynamic feedbacks, as well as thresholds, and their relevance to understanding the nature and dynamics of poverty.

Yet despite a move to a more dynamic multi-dimensional approach to understanding poverty in mainstream development research, and the importance of shocks and interlinked stresses, Boyden and Cooper (2009) assert that 'the resilience concept has not yet been demonstrated as a valid analytical tool for poverty research' (p303). Their assessment is based on resilience understandings from psychology and human development, though there is resonance too with resilience thinking on social ecological systems. For example, they claim that there are some critical challenges and limitation in resilience concepts, regarding it as a 'metaphor'. The disciplinary legacy of resilience in psychology and social work leads to a focus on individuals and overlooks collective and structural aspects of poverty – in effect, it 'depoliticises the project of poverty reduction' (Boyden and Cooper 2008: 295). But the multi-disciplinary and interdisciplinary nature of resilience research means there is no coherent theoretical or conceptual framework – research starts with an assumption of a problem, so has a tendency to be normative. Resilience research also depends on the definition of 'positive adaptation' so that 'risk and resilience are only defined in relation to a problem or positive adaptation definitions' (p297). A final and very important challenge is that resilience cannot be directly observed or measured – hence they refer to it as a 'metaphor'. This issue was touched upon in Chapter 2 in terms of how development agencies are attempting to 'pin down' resilience and devise operational standards and measures. This has been applied especially in fields of vulnerability and food security, as a growing area of applied development studies. But I examine in more detail understandings of chronic poverty and the linkages and synergies – conceptually and practically – between resilience and mainstream development analysis, particularly in the context of environmental and climate change. I counter Boyden and Cooper's assertion, and suggest that the systems perspective of social ecological systems resilience, and the insights from the fourth wave of resilience studies in human development can bring a powerful new lens to understanding the dynamics of poverty and environmental change.

Resilience insights into traps

The persistence of poverty is a major, central and defining concern for development. The focus on chronic poverty explores how individuals, households, communities and even nations remain in poverty, and seeks to identify pathways or trajectories to escape poverty. A key concept which links thinking about poverty with resilience analysis is that of traps. The development literature has long recognised poverty traps, and rigidity traps, and more recently, social ecological traps feature in social ecological systems resilience literature.

A poverty trap is a self-reinforcing mechanism that causes poverty to persist; it results in what has been observed as a vicious cycle of poverty. A poverty trap is about staying poor, not just being poor at any one moment in time (Barrett et al. 2011). Research on poverty in development studies has focused on understanding why, how and under what circumstances certain individuals, households, communities, regions and nations stay in poverty (measured in different ways) and why conventional approaches to poverty alleviation appear not to work. However, as Woolcock (2009) explains, the notion of poverty traps represents the dominant discourse on chronic poverty in international and bilateral development agencies, stemming chiefly from an economic perspective. But the concept has been applied by many different social science perspectives to explain the persistence of poverty at different scales and across generations. He calls for a broader social science theory of poverty traps, widening analysis to include consideration of network isolation (or 'geographical poverty traps', p333), social exclusion and cultural explanations. The *World Development Report 2006* (World Bank 2005) expanded poverty traps to consider 'inequality traps' which represent durable structures of economic, political and social difference which serve to keep poor people poor.

The literature on poverty traps identifies four causal mechanisms that reinforce traps: economic thresholds; dysfunctional institutions; neighbourhood effects; and intergenerational transmission of poverty. Barrett and Swallow (2006) have also analysed how poverty traps occur and interact at multiple scales, developing the notion of fractal poverty traps, in which multiple dynamic equilibria exist simultaneously at multiple (micro, meso and/or macro) scales of analysis and are self-reinforcing through feedback effects. Small adjustments at any one of these levels are unlikely to move the system away from its dominant, stable dynamic equilibrium. This has remarkable resonance with resilience concepts and especially perhaps with the cross-scale dynamics suggested by the panarchy concept.

The social ecological systems resilience literature identifies both poverty traps and rigidity traps. Poverty traps are self-reinforcing feedback loops that keep a social ecological system in persistent poverty. Rigidity traps are self-reinforcing and have high connectivity between institutions and networks of agents with high access and holding of resources. According to Scheffer and Westley (2007: 1) 'feedbacks leading to alternative stable models of behaviour occur on levels varying from the cell and the mind to societies. The tendency to lock into a certain pattern comes at the cost of the ability to adjust to new situations. The resulting rigidity limits the ability of persons, groups and companies to respond to new problems'. We have also developed the notion of a 'gilded trap' where social ecological systems get locked into rigidity traps, which are highly profitable but also highly risky (Steneck et al. 2011).

Carpenter and Brock (2008) use resilience theory and specifically the adaptive cycle to distinguish between a poverty trap and rigidity trap. Traps are defined as 'persistent maladaptive situations' (p9). Rigidity traps occur when institutions become highly connected self-reinforcing and inflexible. This differs in the resilience literature from a poverty trap in which connectedness and resilience are low, and the potential for change is not realised (see Table 6.3). According to Carpenter and Brock, social ecological systems that exist in situations of chronic and recurring disasters such as droughts, occupy poverty traps.

This discussion shows that the resilience and development literatures view poverty traps slightly differently, and indeed Maru and colleagues (2012) helpfully synthesise these different perspectives and the conceptualisations and empirical analyses of traps, isolating poverty traps and rigidity traps. Their key findings are summarised in Table 6.4.

Table 6.3 Poverty and rigidity traps compared

Characteristic	Poverty trap	Rigidity trap
Heterogeneity of entities	High	Low
Network connections	Low	High
Capacity to focus	Low	High
Capacity to explore	High	Low
Average stress	Low	High
Capacity to dissipate stress	High	Low
Analogue in animal physiology	Out of shape	Hypertense

Source: Adapted from Carpenter and Brock 2008

Table 6.4 Characteristics of traps

Trap characteristics and challenges for synthesis	Development literature	Resilience literature
Poverty traps interpretation	Self-reinforcing mechanisms that cause poverty to persist	Maladaptive departures from normal adaptive cycle of system
Poverty trap mechanisms	1 Capital thresholds 2 Dysfunctional institutions, or 3 Neighbourhood effects	A system configuration of low potential, low connectedness, and low resilience that locks a system in a maladaptive state
Rigidity traps	No explicit treatment	A maladaptive configuration of high potential, high connectedness, and high resilience. A social system where members of an organisation and their institutions become highly connected, rigid, and inflexible
Strengths	Detailed, specific, and mainly empirical understanding of traps	Broad, general, and mainly theoretical understanding
Weaknesses	Separate, unconnected explanations Neglect of ecological drivers	Inconsistent treatment of traps Tendency to endogenise causation Limits of biological models

Source: Adapted from Maru et al. 2012

But the concept of traps in the development literature generally omits consideration of underlying ecosystems, and Maru and colleagues highlight this as a major weakness. More recently, the notion of a social ecological trap has developed to synthesise these perspectives. Josh Cinner discusses the prevalence of social ecological traps in reef fisheries (2011). A social ecological trap describes a situation where feedbacks between social and ecological systems lead towards an undesirable state that may be difficult or impossible to reverse. The social and ecological dimensions are interlinked such that social dynamics may reinforce ecological feedbacks. The key social processes which drive the system towards a social ecological trap include – in the case of Cinner's analysis of western Indian Ocean fisheries – missing or weak institutions, poverty and resource use, and specific fishing techniques. It is allied, but not identical to a poverty trap; indeed poverty traps become part of the drivers of the social ecological trap, where persistent poverty causes over-fishing and use of destructive gears (see Figure 6.1).

Boonstra and de Boer (2014) view social ecological traps as a concept to explain the rigidity of social and ecological processes that simultaneously produce environmental degradation and livelihood impoverishment. In reviewing studies which use the trap metaphor, Boonstra and de Boer remark that from a historical sociological perspective, social ecological traps are clearly path dependent processes, causally produced through a conjunction of events. They emphasise traps as a process rather than a condition. They claim that the concept of traps has the potential to account for historical dynamics of collective action, but that in most studies of traps, history is treated as a contextual or background variable. In comparing historical sequences of four case studies, they are able to show the causal importance of the conjunction of social and ecological events. These manifest as 'critical junctures' that subsequently trigger the persistence of specific social ecological dynamics – in other words they propel the system deeper into the trap itself. One of the examples they cite, the 'gilded trap' of the lobster fishery in Maine, was the subject of analysis with a group of scholars from the Resilience Alliance (Steneck et al. 2011). The antecedents of the trap were the historical over-fishing during the twentieth century – propelled by high prices, improved technology and open access institutions which propels the systems towards higher catches, more sophisticated vessels, and economic dependence. Critical junctures occurred in the 1980s with new fisheries policies and exclusionary property rights, but these served to enable local fishers to maximise their profits rather than take action to conserve stocks. As other fisheries collapsed, social and ecological feedbacks drove the specialisation of the lobster fishery, developing the 'trap' – a virtual lobster monoculture and a regional economy dependent on one lucrative high value product. This 'gilded' trap is profitable, but perhaps vulnerable, for example, to disease, and communities have developed few livelihood alternatives. Boonstra and de Boer's other example, the dryland poverty trap in Tanzania reports a land degradation-impoverishment cycle which highlights the pre-conditions, including low productivity land and economic marginalisation, and stressors such as increasing populations, changes in land use practices and increased sedentarisation of livestock keepers, and triggers such as persistent drought and crop failures which resulted in self-reinforcing poverty traps for dryland farmers.

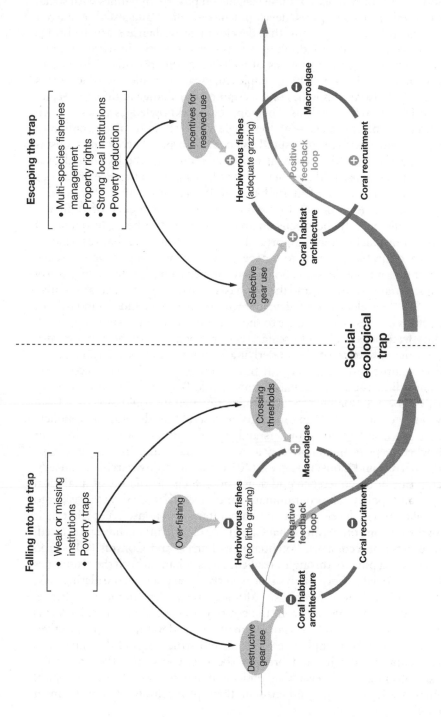

Figure 6.1 An heuristic model of a social ecological trap in coral reef fisheries

Source: Cinner 2011

Another form of synthesis of these insights on poverty dynamics with social ecological resilience concepts is 'development resilience', suggested by Barrett and Constas (2012). They claim that development resilience starts to bridge the divide between resilience thinking and poverty analysis. In foregrounding risk in understanding the dynamics of well-being, this provides purchase in trying to understand not only how people can escape chronic and other forms of poverty, but also brings issues of uncertainty and change at different scales more explicitly into this analysis. Their approach acknowledges shocks – such as ill health or bereavement – as the single greatest cause of descents into poverty, combined with the uninsured risk of catastrophe. Barrett and Constas (2014: 16) assert that 'development resilience represents the likelihood over time of a person, household or other unit not being poor in the face of various stressors and in the wake of myriad shocks. If and only if that likelihood is high, then a unit is resilient'. In seeking to develop a theory of resilience against chronic poverty they focus on an individual's well-being or quality of life within a broader system. Importantly, they acknowledge that resilience is not always desirable and that to escape poverty people may have to escape from their present (stable) state. So the tension between persistence of poverty and transformation is noted, and the emphasis is on the need for transformative change. The key elements of development resilience include standards of living, effects of stressors, response to shocks, and a dynamic systems approach including feedbacks and multi-scalar reinforcements, and the authors also recognise the resilience of the underlying resource base as instrumental in resilience against chronic poverty. It thus points to links between ecological and human dimensions in a dynamic framework. Barrett and Constas call for resilience to be a normative property, suggesting that it should directly inform policy and action (2014). In order to be useful for development policy, resilience can also help us to understand the dynamics of poverty – in other words what keeps people in poverty as well as what will help them to escape poverty. Barrett and Constas suggest (2014) that as an overarching concept, resilience can bring coherence and a pro-poor focus to the set of analytical concepts and policy of programmatic initiatives that deal with risk and vulnerability and in development and humanitarian responses. Yet what their current theorising omits, and what Cinner (2011) and others have integrated, are the ecological feedbacks. Potentially, as Barrett and Constas (2014) too suggest, viewing poverty through a resilience lens enhances both the economics and broader social science analysis of poverty by explicitly considering risk, dynamics and ecological feedbacks. Although their 'development resilience' integrates ideas on resilience from the ecology and social ecological systems literature, it does not include consideration of human development perspectives on resilience. So it too emphasises systems and structure and de-emphasises agency, adaptive capacity and broader social sciences. But they recognise these shortcomings and posit that understanding the social and ecological dynamics needs to be strengthened in their proposition of 'development resilience'.

As Barrett and Constas (2014) show, and as Josh Cinner (2011) has cogently explained using a small-scale fisheries example (Figure 6.1), the feedbacks between social and ecological components of the system become critically important. *Poverty dynamics* (Addison et al. 2009), targeted to a development scholarly audience, is interesting for what it does and does not say about resilience ideas and about dynamics. Despite its orientation towards dynamics and its discussion of multi-dimensional poverty and the ways in which intersecting stress and crises affect poverty and its persistence, the book makes no reference to environmental change or climate change. Similarly, of the three *Chronic Poverty* (CPRC) reports, 2004–5, 2008–9 and 2014–15, only the latest incorporates environmental change and climate change into the analysis, with additional mention in the 2008–9 report of vulnerability and climate variability in sub-Saharan Africa. The latest report gives greater emphasis and analytical depth to understanding the roles of shocks and disasters, including environmental disasters, than previous reports, but generally these observations show the separation of mainstream international development and environmental change research and scholarship.

Conventional understanding of poverty traps has limitations – it is based in the main on economic analysis and does not adequately deal with the sociological, historical and cultural – it is too narrow. Resilience concepts such as poverty traps and rigidity traps might then explain why people get trapped in poverty and why poverty becomes self-reinforcing, and particularly emphasises the role of feedbacks. This highlights possible multiple states and moves away from equilibrium focus, and it also draws attention to the importance of thresholds. Here the concept of hysteresis from resilience is important. This essentially means that it is much more difficult to escape from poverty than to fall into it; the path out is different to the path in. Although such a move is not completely irreversible, it takes a much bigger effort to go back up and return to the previous state.[3] So the extent to which this can inform the transition out of poverty is limited.

Escaping traps and transformation out of poverty

Eradicating chronic poverty and escaping social or poverty traps requires major changes, or transformations, to overturn the structures and processes, and to override 'lock-in'. The need for transformative change, for social transformation, has been a major focus of development studies and a central objective of international development. Only relatively recently has this recognition of the need for fundamental change to structures and processes become central to the environmental change literature. However, in the past few years, the environmental change and climate change fields have more explicitly advocated for transformational change in order to both successfully adapt to, and also to mitigate, environmental change. We reviewed this shift in thinking in a chapter in the *World Social Science Report 2013*, which emphatically sought to link the complexity and urgency of global environmental change and social transformation (Brown et al. 2013).

The key argument here is that although much of the literature on environmental change makes the case for transformations in response particularly to climate change, there are few empirical examples or cases to build on. Likewise, the resilience literature on social ecological systems uses a few, iconic examples to develop empirical evidence of transformation. So in these fields much less is known about how transformation happens than in development studies and social science more broadly, so this is where resilience thinking can be informed by these other literatures. Yet even within fields, there is relatively little agreement on key concepts, and apparently little consensus on exactly what constitutes a transformation (Moore et al. 2014; Brown et al. 2013). A failure to move beyond the status quo or business as usual has been identified as a problem with resilience approaches. This may be related perhaps to the roots of the term as 'bouncing back' – many applications and some scholarship around resilience links the property very firmly to the ability to bounce forward and to radically change. This section outlines ideas about social transformation from development studies and the broader social science literature, before then examining how the social ecological systems resilience field understands and approaches transformations. The chapter concludes by synthesising the elements of these fields that are especially relevant for understanding transformations applied to international development and environmental change.

Contemporary literature on poverty explicitly recognises the necessity of transformation to overturn chronic poverty and to enable people to escape poverty traps. The *Chronic Poverty Report* (CPRC 2008) argues that there are five reasons for the persistence of poverty traps: insecurity, limited citizenship, spatial disadvantage, social discrimination and poor work opportunities. They suggest five strategies to escape these poverty traps: anti-discrimination and gender empowerment; building individual and collective assets; strategic urbanisation and migration; social protection; and public services for the hard-to-reach. Without these measures, the poorest will remain in persistent poverty. Economic growth is important, and growth in sectors of the economy where the poor are concentrated – such as agriculture – is necessary but not sufficient. Growth is only effective in helping the poor escape poverty traps when it opens spaces for social change and political momentum, for example, redistribution, regulation of markets, investment and greater financial accountability to enable the poor to benefit. This will only happen if the poor have greater agency, have routes out of oppressive social relationships and political voice. This requires social transformation. The policies required to make way for this social transformation to eradicate chronic poverty as suggested in the report are shown in Table 6.5. This builds a comprehensive picture of how social transformation might be brought about in different poverty and national

Table 6.5 National policies for transformative societal change

Policies	Chronically deprived countries (CDCs)	Recently improving CDCs	Partially chronically deprived countries	Consistent improvers
Anti-discrimination	Specific legislation to promote inclusion of minorities, to ensure political stability	Legislation on gender equality; legislation or other measures (e.g. minimum wages) to control discrimination in labour markets	Comprehensive constitutional agreements Legislation on child labour to tighten labour markets	Broader regulation of labour markets
Urbanisation	Develop national urbanisation policy and infrastructure investments – getting the most out of urbanisation for the poor as a whole	City policies on pro-poor (labour-intensive) growth and gender equality	Quality mass urban services to break adverse incorporation	Quality mass urban services to break adverse incorporation
Migration	Migrant support programmes; public information to counter negative stereotypes	Migrant support programmes; public information to counter negative stereotypes; remove restrictions on mobility	Open up trans-national movement; remove restrictions on labour mobility	Implement migrant rights and entitlements
Demographic transition	Foster demand and improve supply of reproductive health services (RHS)	Foster demand and improve supply of RHS	Sustain demand through girls' education, universal access to RHS	Sustain demand through girls' education, universal access to RHS
Post-primary education (PPE)	Universal PPE (including technical/vocational)	Universal, compulsory PPE; widespread scholarships for poor children (girls) in secondary and tertiary education	Free and compulsory PPE	Support for tertiary education, including extensive scholarships for the poor
Social movements	Public authority strengthened and civil and political rights protected, as a means of encouraging citizen engagement and reducing transaction costs for citizenship participation	Social movements focus on asset redistribution and social protection	Direct support to actors within social movements that maintain a focus on the poorest	Coalition-building between social movements and political institutions

Source: Chronic Poverty Research Centre (2008): 85

contexts. Again it moves away from a 'one-size-fits-all' approach and a central focus on growth, and clearly identifies the structural changes necessary at a national scale to make development activities effective for the poorest people. It gives a very detailed and comprehensive view of what is necessary for social transformation to eradicate poverty.

In a recent chapter for the *World Social Science Report 2013* (WSSR) (Brown et al. 2013) we present a review of what is known about transformations from across the social sciences that could inform calls for transformation in response to global environmental change. We looked across the social sciences for common approaches to understanding transformations: the meanings are quite diverse, and obviously specific to contexts and fields, but they share a commonality in terms of transformation representing a fundamental shift in structure, functions or understandings of how the world works. But in social sciences, transformation has ambiguities and multiple meanings, implying shifts in both structure (changes to institutions, cultures) and agency (empowering people to envisage and implement alternative pathways). Very often, discussions of transformations conflate individual, collective and broader system or regime-scale change.

We examine how transformations had been presented in the environmental change literature, and where the empirical evidence lay. Some of the examples we cite are shown in Table 6.6. We analyse these in terms of the scale they occurred at (does transformation always have to be large-scale?), who the key actors are and the degree of anticipation. Some transformations are deliberately induced as a response to experienced or anticipated environmental change – so in effect they are extreme forms of adaptation. They highlight the distinctions made in the literature on transformations between deliberate and unplanned transformations, and anticipatory and reactive transformation (echoing typologies in the adaptation literature). They highlight different triggers and different agents, as well as different scales of transformation.

These distinctions are discussed in most literatures around transformation. But notably, we find a very rapid increase in the language of transformation in debates around global environmental change and societal responses (O'Brien 2012; Hackmann & St Clair 2012). This field argues first that global environmental change will enforce radical, unplanned and detrimental transformation especially through impacts of climate change. Second, there is a normative imperative for planned and profound transformation especially of energy and consumption to avoid the worst impacts of these changes, and to implement sustainability (Kates et al. 2012).

As we started to examine various examples of social transformations we found that the drivers and catalysts of transformation act at multiple spatial and temporal scales and may take the form of gradual shifts or fast changes, and may be punctuated by surprises or episodic events, such as a series of decisions and shifts in practices by farmers inducing and responding to

Table 6.6 Examples of transformation in response to environmental change

Documented example	What transformed?	Key characteristics scale, key actors, degree of anticipation
Great Barrier Reef, Australia	The transformation saw the focus of governance shift from protection of selected individual reefs to stewardship of the larger-scale seascape.	A transformation process was induced because of increased pressure on the Great Barrier Reef (from terrestrial runoff, over-harvesting, and climate change). Reformulation of governance was supported by changes in legislation.
Flood management, Netherlands	Coastal defence and riverine flood abatement and water supply are transformational because of their enlarged scale, intensity and integrated combinations of adaptations, which includes novel approaches, like artificial islands, evacuation of some areas, as well as new institutions and funding mechanisms.	Planned interventions through government and in response to both experienced and anticipated sea level rise and flooding.
Energy systems, Germany	New technologies, regulatory regimes, management styles, marketing strategies and environmental priorities have emerged, dramatically reconfiguring patterns of governance within cities and regions such as Berlin.	Transformation has been triggered by new technology, economic conditions and legal frameworks.
Energy systems, China	Energy generation has been transformed in Rizhao a coastal city of nearly three million people in northern China. There has been rapid and widespread adoption of renewables: e.g. 99% of households in the central districts use solar water heaters, and public infrastructure is powered by photovoltaic cells.	State-led, rapid and intentional wide-scale transformation.
Transformations in prehistoric American societies	Archaeological research shows that two factors were closely associated with rapid and severe transformations of social ecological systems in prehistoric American societies: limitations in social and materials diversity; and investments in infrastructure that limited options.	Comparison of the Hohokam, Mimbres, and Zuni societies explore trade-offs between short-term efficiency and long-term persistence. The Hohokam system was heavily capitalised toward efficiency for improved productivity and collapsed most dramatically. The Zuni who did not invest in expensive and long-lasting infrastructure for farming, and maintained extensive social networks and distant linkages experienced less dramatic transformations.

Source:: Adapted from Brown et al. 2013

larger landscape changes and agricultural sector or market changes. What is especially evident is that often the interplay between fast and slow drivers operating at global, national and sub-national scales results in unpredictable and 'messy' transformative processes. Transformation is seldom a neat 'flip' from one state to another. In most instances many elements of pre-transformed systems linger on as memory in the new system, ready to revive themselves when a combination of events create conducive conditions. This is exemplified by the example of the multiple dimensions in South Africa's transformation to peaceful democracy, which we used in the WSSR chapter (Brown et al. 2013). Thus, rapid, catalytic changes at a national and sub-national scale included the Soweto uprising; unbanning of liberation movements; democratic elections; Hani's assassination; Botha's Rubicon speech, and his replacement by de Klerk. These happened against a backdrop of global slow and faster changes, such as the fall of the Berlin Wall, collapse of the Soviet Union and political change in Europe, and gradual changes like the impacts of sanction and rise in electronic communications. National slow changes included the civil resistance to apartheid, the preparation of the ANC in exile and economic change. This gives a sense of the multi-dimensional and historic nature of transformation that is largely absent from analyses in environmental change, but much more evident when we think of large-scale social transformations such as the Arab Spring or the transition from state to market economies.

Transformation is thus rarely a discrete and tidy event. It may indeed be a process which is triggered by a specific event but which develops somewhat messily over time and space. In many respects, this makes it more difficult to say when a series of changes constitute a transformation. For example Tiffen et al.'s (1994) documentation of landscape-scale transformation in upland Kenya can be seen to be the product of a series of discrete changes at different scales: individuals' migration decisions; farm-scale land use decisions; changes in markets and information; and changes in government infrastructure. Olsson et al.'s (2006) analysis of five case studies of transformation of governance of social ecological systems shows how transformation involves shifts in social features such as perceptions, network configurations, social co-ordination, and institutional arrangements and organisational structures, and how it may be triggered by dramatic events. Folke et al. (2010) cite the example of the Goulburn watershed in southeast Australia to illustrate the cross-scale interplay and transactional relationship between scales and sites. Deliberate transformational change at the scale of the whole catchment, and of all its component parts at the same time, is likely to be too costly, undesirable or socially unacceptable. Transformational changes at lower scales, in a sequential way, can lead to feedback effects at the catchment scale, partly a learning process, and facilitate eventual catchment scale transformational change.

A critical issue for environmental change is how we distinguish between adaptation and transformation. This has led some authors to develop the notion of 'transformational adaptation' (O'Neill & Handmer 2012); others see transformation at the end of a continuum of adaptation (Schoon et al. 2011), or as a type of adaptation (Pelling 2010); and others view transformation as quite distinct from adaptation (Marshall et al. 2012). But for Nelson et al. (2007) transformation is distinguished from incremental adjustments by its outcome – transformation involves crossing a social or ecological threshold and creating a fundamentally new social ecological system. In the Nelson et al. (2007) analysis (see Chapter 5), agricultural collapse in Jordan and a shift from agriculture to tourism in Arizona, USA, are cited as examples of unplanned and planned transformation.

The role of resilience concepts in this discussion of transformation is ambiguous. As already discussed, resilience is often viewed as inherently conservative and the central focus on bouncing back works against more profound change. However, as noted in Chapter 3, more recent writings on resilience in social ecological systems signal a realignment – indeed a redefinition – of resilience linked to profound change and to transformation. Although the edited volume, *Panarchy* (Gunderson & Holling 2002) and Walker et al.'s 2004 paper theorised transformation and resilience, recent literature has re-affirmed this emphasis with frequent linking of resilience and transformation. For example Folke et al. (2010) assert that adaptation and transformation are essential to maintain resilience, that 'the very dynamics between periods of abrupt and gradual change and the capacity to adapt and transform for persistence are at the core of resilience of social ecological systems' (p1). The key to this – apparently counterintuitive – relationship between transformation and adaptation is cross-scale dynamics, so that according to Folke et al. (2010: 3) active transformation is 'the deliberate initiation of a phased introduction of one or more new state variables at lower scales, while maintaining the resilience of the system at higher scale as transformational change proceeds'. For example, the transformation of transport, energy systems or human behaviour is necessary to maintain resilience of the global climate system. The social ecological systems resilience literature generally acknowledges that transformational change involves not just a shift in state variables (for example, those which control the underlying state of the system such as the loss of summer sea ice which Folke et al. refer to may transform the geopolitical and economic feedbacks among Arctic nations), but also shifts in perception and meaning, patterns of interaction among actors including leadership and political and power relations and institutional arrangements (Folke et al. 2010). Thus Chapin et al. (2009) refer to transformation as a fundamental change in a social ecological system resulting in different controls over system properties, new ways of making a living and often changes in scales of crucial feedbacks. Adjustments occur at all (and interlinked) scales: for individuals, society, institutions, technology, economy and ecology; and may involve changes to practices, lifestyles, power relations, norms and values. There is often an emphasis on learning, and transformation necessitates a commitment to innovation, novelty and diversity in order to imagine alternatives and possible futures (Schoon et al. 2011). Another collection of papers discusses the role and scope for agency, innovation and novelty within resilience framing for understanding transformation, adopting language and concepts from transitions literature (Westley et al. 2011; Loorbach 2010; Haxeltine & Seyfang 2009). Olsson and colleagues used transition theoretical constructs to analyse transformations in management of social ecological systems, for example, in Swedish wetlands and the Great Barrier Reef Marine Park (Olsson et al. 2006).

But there is on-going debate, especially in the environmental change literature, about the relationship between transformational change and resilience. Some analysts contend that resilience is quite distinct from and cannot support transformational change. In line with the views of resilience as conservative, Pelling's work on climate change adaptation (2010) sees resilience approaches as quite distinct from transition or transformation. Indeed some recent writings separate adaptive resilience and transformative resilience (Christmann et al. 2012; Wilson et al. 2013). But this topic remains empirically rather sparse (important exceptions being Marshall et al. 2012 and Park et al. 2012). There is no single agreed definition or understanding of transformation, and many normative assumptions abound, not least about the assumed desirability of transformational change – echoing normative stances on resilience (Brown et al. 2013). Here, Carpenter and Brock's (2008: 1) contention that resilience 'explains how transformation and persistence work together, allowing living systems to assimilate disturbances, innovation and change, whilst at the same time maintaining structures and processes' – in other words it provides insights into the *interplay* of transformation and persistence – is helpful if we think about how this applies to poverty.

A recent paper by Moore et al. (2014) integrates insights from the social ecological systems resilience literature on transformations with concepts applied in studying social innovations, transition management and social movements. Acknowledging that the social ecological systems understanding of transformations downplays issues such as power, they seek to use these other literatures to develop a better framework to understand the social dimensions of transformations in social ecological systems. They ask, what changes in transformation? and then clarify the ecological and social dimensions that constitute transformations (see Box 6.1).

They recognise transformation as distinct from adaptation, but highlight that the course of transformation involves changes in both structure and agency, reflecting diverse social science understandings. Although transformation is unpredictable, it is not random, in line with complex adaptive systems thinking. But across the different knowledge domains there are disagreements, for example, about the degree to which an endpoint is identified and agreed. Their analysis breaks transformation down into a series of processes and sub-processes. This highlights the dynamic between change and finding stability in a new state. But the emphasis here is clearly

---

Box 6.1 Transformation: what changes?

Ecological transformations include:	Social transformations include:
Natural capital processes, functions and specifies configurations	Norms, values and beliefs
Ecosystem services and feedbacks	Rules, practices, regulations
	Distributions and flow of power, authority and resources

Source: Author's own, synthesised from Moore et al. 2014.

on the social and power dynamics of change, and Moore et al. (2014) emphasise that any durable transformation requires altering the dominant structures of power and embedding newly configured social and ecological elements and feedbacks within new institutions as the new trajectory gains power.

The international development, environmental change and social ecological systems literatures converge on the need for the transformation of society to address sustainable development and global change. This is reflected strongly in contemporary policy documents, as well as international science organisations (e.g. Hackmann and St Clair for the International Social Science Council 2012; German Advisory Council on Global Change n.d.). Yet the analysis in Chapter 2 shows that despite the rhetoric of change and resilience, many of the policies and strategies suggested represent incremental change. We know that powerful actors generally are compelled to maintain the status quo and to resist fundamental change, and of course that transformations are likely to produce 'winners and losers'. I've shown here how the different fields and literatures on transformation might be mutually beneficial. There are strong indications that deliberate transformation is necessary especially when current ecological, social and economic conditions are untenable or undesirable. Karen O'Brien (2012), amongst others, has written most eloquently about this. But the crucial questions for social science and for international development are about who decides and how is transformation negotiated and enacted – in fact how can we deliberate transformation? The synthesis by Moore et al. (2014) points to some of the core issues that arise when we start to understand the *process* of transformation. I turn next to what a transformative approach integrating insights from resilience and social ecological systems, and the different strands of social science, might mean for international development policies and priorities.

Transforming development

As Leach et al. (2010) have very cogently articulated, a science of social and ecological dynamics must be better linked with core concerns for social justice to better inform sustainable development. My aim is to explore how a socially and politically informed perspective which uses social ecological systems *and* human development resilience concepts might suggest the types of transformations necessary in international development to overcome enduring and persistent poverty in face of environmental and global changes. So far I've presented areas of science that combine development thinking and resilience concepts to understand the dynamic interplay between persistence and transformation, as Carpenter and Brock (2008) suggest, to inform poverty and its alleviation. But what does this mean for development policy and actions; how can these concepts be applied? These implications are discussed in this final part of the chapter. I present a more comprehensive re-visioning of resilience for international development in the next and final chapter.

In a recent edited volume, Biggs et al. (2015) distil key findings from a broad resilience literature to test seven principles that underpin building resilience in social ecological systems. They review the different social and ecological factors that have been identified in the (broad) social ecological systems literature to enhance resilience, and synthesise seven principles that are considered crucial to build resilience. These are: 1) maintain diversity and redundancy; 2) manage connectivity; 3) manage slow variables and feedbacks; 4) foster complex adaptive systems thinking; 5) encourage learning; 6) broaden participation; and 7) promote polycentric governance systems. They discuss how these principles can be practically applied, and they highlight the importance of context and the crucial influence of processes – of how the principles are applied and by whom. But their discussions are about social ecological systems in general. Table 6.7 summarises some of the ways in which these principles apply, or have been applied, in international development. This table links these seven resilience principles, broad and generic as they are, to some of the key themes and strategies followed and prescribed in international development.

This finds some resonance between the resilience principles and selected strategies that have been promoted in international development. But the applications are often piecemeal. As the analysis in Chapter 2 demonstrates, resilience terminology and concepts have been widely adopted by international development and environmental agencies in response to perceived global threats, particularly of climate change. But in many cases this has signalled a shift in rhetoric and only relatively incremental changes in practice. The extent to which the applications identified in Table 6.7 represent real concrete changes on the ground is far less clear. Much work has discussed the 'mainstreaming' of different concepts (gender, participation, basic needs) in development, and the adoption of 'buzzwords', indicating that the opportunity for more substantial or paradigmatic shift is more limited.

Table 6.7 Resilience principles and their application to international development

Principle	Explanation	Application in development field
1 Diversity and redundancy	Systems with many different components are generally more resilient; response diversity especially important	Well established in field on livelihood diversity and diversification, and farmer/agrodiversity, but also governance diversity e.g. institutional diversity
2 Connectivity	Connectivity refers to the structure and strength with which resources, species or actors disperse, migrate or interact across different domains in social ecological systems (SES). But connectivity has both positive and negative effects	Role of social institutions and networks; role of key actors, brokers or leaders; new information technologies; hyper-connected and centralised systems (e.g. energy) may be at risk
3 Slow variables and feedbacks	Slow variables and feedbacks are important drivers of change and may help to keep a SES 'configured', but often difficult to detect and monitor	Cultural norms and social change e.g. role and status of women; role of traditional institutions; importance of legal frameworks
4 Complex adaptive systems	Acknowledging uncertainties and interdependencies	Urged in humanitarian aid and disasters field; move away from 'sectoral' approaches and technical fixes; incorporating risk into investment and insurance strategies
5 Learning	Earning and experimentation through adaptive and collaborative management	Long tradition of capacity building and participatory learning; integration of 'traditional', indigenous technical, traditional ecological and other forms of knowledge, implemented through e.g. farmer field schools
6 Participation	Involving diversity of stakeholders to build legitimacy and trust, expand opportunities for sharing diverse knowledge and detecting and managing change	Key theme in development, 'mainstreamed' since 1980s; sometimes queried and challenged for not taking account of power dynamics and reinforcing asymmetries; new forms of multi-stakeholder fora
7 Polycentric governance	Improves connectivity and learning; able to respond more readily to change ad disturbance	Calls for multi-scale governance, scaling up, decentralisation

Source: Developed from seven principles suggested by Biggs et al. 2015

To what extent then does development itself need to transform to meet the challenges of climate and other global changes, and how might it transform? In 2009, I published an article with Nick Brooks and Natasha Grist in the journal *Development Policy Review* (Brooks et al. 2009) that argues for a fundamental shift in development thinking to take account of the challenges thrown up by environmental change generally and climate change specifically. These challenges for development include those of uncertainty, non-linear change and multiple interacting stressors. We analysed historical records to show how societies in the past had responded to major climatic change. The lessons from the past highlight the need for precautionary action; to act in anticipation of change and in the face of uncertainty; to work at multiple scales; to develop institutions which will adapt; and to incorporate learning. All of these suggest that a radical rethinking of how development is conceived and implemented is required to deal with the scale of change currently faced by global society. In particular, we argue, there is a need to move beyond current global climate change discourses that emphasise managerial and technocratic solutions, and to seek radical departures from the dominant developmentalist paradigm and current conceptions of sustainable development. Development will need to be based on systems and approaches that can accommodate large changes in mean climatic and environmental conditions, enhanced climatic variability over a range of timescales, and (in many parts of the world) a high degree of uncertainty regarding how climate will evolve in the near, medium and long term. In addition, development has to grapple with very specific consequences of climate change. These include, in different contexts, the loss of productive land, the inundation of coastal settlements, and systemic changes in landscapes, ecosystems and resource availability, as well as changes in the nature, distribution, frequency and severity of climate-related hazards like drought, heat waves, severe storms and rainfall patterns. It is now widely acknowledged that these changes will make development along business-as-usual lines impossible and infeasible in many parts of the world.

We distilled the implications of these factors for different sectors, or different spheres of development. Based on this analysis and integrating concepts from social ecological systems and human development resilience, Table 6.8 presents how a new approach might be forged for development in the face of climate change, and highlights how this contrasts with current development strategies and paradigms. Critically it demands a move away from economic growth as a core strategy and overarching goal, which still lies at the heart of current initiatives and responses to global (environmental) change, such as Green Growth or Climate Resilient Development or Pro-Poor Growth, and Sustainable Intensification. Despite attempts to bring together goals of poverty eradication and address global change and climate change in the Sustainable Development Goals, commentators such as Pogge and Sengupta (2015) suggest that the economic, political and institutional transformations necessary to meet these goals and which drive global

Table 6.8 International development in the face of climate change

	Current paradigm	*New approach*	*Policy implications*
Agriculture	'Sustainable intensification': maximise agricultural yields; GM crops	Minimise risk; promote resilience; address variability	Addressing and re-balancing equity through restructuring of land tenure and other access issues; ecosystem services approaches; promote agrodiversity and livelihood diversity
Economy and development	Maximise economic growth; increasing concerns with equity and 'pro-poor growth'	Minimise environmental and social impacts; emphasis on equity, multi-dimensional poverty and well-being	Sustainable or 'resilient' development to become principal models of growth – myriad of regional, national and local variations; tailor development to the situation
Institutional approach	Managerialism; technocratic approach to resolving climate problems pump-priming the solution (e.g. African Green Revolution, Green New Deal)	Multiple approaches embracing social, political and capacity elements of development and resilience first, before bringing in technocratic solutions; horizontal and vertical integration of governance	Embrace diversity within overarching international targets; strong global environmental governance essential to ensure well-being of the poorest through UNFCCC; research and funding on climate change to be focused on the development reality experienced first, building up over time
Sustainability	Environment treated as an externality; certain systems and modes of production viewed as optimal, regardless of environmental contexts, resulting in tendency for development to marginalise diversity and adopt universalist approaches	Environment prioritised in calculations; development built around environmental constraints and opportunities	Take on concepts of non-substitutable environmental capital – major implications for land use and management worldwide as non-renewables and renewables become valued; promote systems designed specifically to function under extant (and anticipated) environmental conditions, appropriate to local and regional circumstances

	Current paradigm	New approach	Policy implications
Environment	Assumption of static environment; single equilibrium; manage for stability	Dynamic environment, multiple equilibria; manage for change	Build in leeway in the use of the environment; science focus on tipping points and limits; precautionary approach
Social development and well-being	Subsidiary to economic growth; flowing from economic growth	Social indicators and non-economic quality of life are foremost	New approaches to promoting well-being; development focuses on delivering food security, health, clean water and risk management rather than economic growth
Urbanisation	Seek to provide services for growing population	Promote sustaining agricultural lifestyles more strongly; promote multiple, resilient livelihood approaches based in rural and urban places	New approaches to urban planning, considering evolution of exposure to climate-related hazards over long (i.e. multi-decadal) timescales Increased urban food production Understanding cities as 'urban systems'
Climate change	Make development 'resilient' to climate change; 'Climate Resilient Development'; mainstreaming adaptation; 'Climate Smart Agriculture' focus on national level plans	Low carbon development and innovation; understanding climate change in concert with other stressors	Avoid maladaptation; promote resilience at multiple scales; address cross-sectoral and cross-scale issues; address roots of vulnerability; strengthen capacities at all levels

Source: Adapted from Brooks et al. 2009

inequality and unsustainability will not be delivered by the current suggested agenda. In particular, contradictions and inconsistencies, and failure to address underlying causes of climate change and to effectively link this with achievement of other goals, mean the goals are unlikely to be workable or achievable without significant institutional reforms.

As this table suggests, resilience ideas and concepts can provide some guidance for transformation of development in the era of global environmental change. The discussion in this chapter has shown how key concepts are shared – or have significant similarities – across the development and resilience literatures. Recent analysis of poverty suggests that a more dynamic perspective is required, and a move away from single disciplinary focus to multi-dimension understandings of poverty.

Eliminating chronic poverty is and will remain a central thrust and focus of development. Analytically this means understanding why people stay in poverty – hence a concentration on the notion of 'traps'. Different perspectives from social ecological systems research and development show some similarities, and there is a suggestion that bringing resilience insights into understanding poverty traps could be fruitful. Barrett and Constas (2014) in particular suggest the need for theoretical advance in 'development resilience'. At present this needs more empirical testing, but it also requires integrating feedbacks from ecological systems, along the lines of Josh Cinner's (2011) notion of social ecological traps which brings together concepts from poverty and resilience perspectives.

Second, the discussion in this chapter suggests transformation – or structural and systemic change – is necessary to address chronic poverty. This becomes another meeting point for contemporary science of resilience and development. Whilst social transformation lies at the heart of development thinking, the literature on transformation in resilience is still relatively immature, and not empirically informed. I suggest ways in which resilience concepts could be used to inform transformative development in the face of environmental change and other challenges. But what aspects of resilience – what concepts, analytical perspectives, which dimensions, meanings and views – are likely to inform and support this kind of development? The next and final chapter presents a re-visioning of resilience that will more directly and appropriately speak to and provide analytical traction for development and transformation in the era of global change.

Notes

1   www.espa.ac.uk/files/espa/ESPA-Poverty-Framework.pdf.
2   www.chronicpoverty.org.
3   See Glossary.

References

Addison T, Hulme D and Kanbur R (2009) Poverty Dynamics. In: Addison T, Hulme D, and Kanbur R (eds), *Poverty dynamics: Interdisciplinary perspectives*, Oxford: Oxford University Press, 3–26.

Barrett CB and Constas MA (2012) Resilience to avoid and escape chronic poverty: Theoretical foundations and measurement principles. In: *Presentation at CARE USA's Washington, DC, Roundtable on Resilience, 2012.*

Barrett C and Constas M (2014) Toward a theory of resilience for international development applications. *Proceedings of the National Academy of Science,* 111(40), 14625–30.

Barrett C and Swallow B (2006) Fractal poverty traps. *World Development*, 34(1), 1–15.

Barrett C, Travis A and Dasgupta P (2011) On biodiversity conservation and poverty traps. *Proceedings of the National Academy of Science*, 108(34), 13907–12.

Béné C, Evans L, Mills D, et al. (2011) Testing resilience thinking in a poverty context: Experience from the Niger River basin. *Global Environmental Change*, 21(4), 1173–1184.

Biggs R, Schlüter M and Schoon M (eds) (2015) *Principles for building resilience: Sustaining ecosystem services in social ecological systems*. Cambridge: Cambridge University Press.

Bird K and Higgins K (2011) *Stopping the intergenerational transmission of poverty: Research highlights and policy recommendations*. Available from: http://www. chronicpoverty.org/uploads/publication_files/WP214 Bird and Higgins.pdf.

Boonstra WJ and de Boer FW (2014) The historical dynamics of social ecological traps. *Ambio*, 43(3), 260–74.

Boyden J and Cooper E (2009) Questioning the power of resilience: Are children up to the task of disrupting the transmission of poverty. In: Addison T, Hulme D, and Kanbur R (eds), *Poverty dynamics: Interdisciplinary perspectives*, Oxford: Oxford University Press, 289–308.

Brooks N, Grist N and Brown K (2009) Development futures in the context of climate change: Challenging the present and learning from the past. *Development Policy Review*, 27(6), 741–65.

Brown K, O'Neill S and Fabricius C (2013) Transformation: Social science perspectives. In: Hackmann and St Clair (eds) *World Social Science Report 2013 – Changing Global Environments: Transformative Impact of Social Sciences*. Paris: UNESCO, Chapter 9, 98–104.

Carpenter SR and Brock WA (2008) Adaptive capacity and traps. *Ecology and Society* 13(2): 40.

Chapin FS, Kofinas GP and Folke C (2009) A framework for understanding change. In: Chapin F, Kofinas G, and Folke C (eds), *Principles of ecosystem stewardship*, New York: Springer, 3–28.

Christmann G, Ibert O, Kilper H and Moss T (2012) *Vulnerability and resilience in a socio-spatial perspective: towards a theoretical framework. Working Paper 45*, Available from: www.irs-net.de/download/wp_vulnerability.pdf (accessed 14 May 2015).

Chronic Poverty Research Centre (CPRC) (2005) *The chronic poverty report 2004–2005: Social protection*. Available from: http://bit.ly/1e2ipaK.

Chronic Poverty Research Centre (CPRC) (2008) *The chronic poverty report 2008–09: Escaping poverty traps*. Available from: www.chronicpoverty.org/uploads/publication_files/CPR2_ReportFull.pdf.

Chronic Poverty Research Centre (CPRC) (2014) *The chronic poverty report 2014–2015: The road to zero extreme poverty*.

Cinner JE (2011) Social ecological traps in reef fisheries. *Global Environmental Change*, 21(3), 835–39.

Davis P (2009) Poverty in time: Exploring poverty dynamics from life history interviews, in Bangladesh. In: Addison T, Hulme D, and Kanbur R (eds), *Poverty dynamics: Interdisciplinary perspectives*, Oxford: Oxford University Press, 154–82.

DFID (2011) *Defining disaster resilience: DFID approach paper*. London: Department for International Development.

Ellis F (2000) *Rural livelihoods and diversity in developing countries*. Oxford: Oxford University Press.

ESPA (2012) *Conceptual framework: poverty*. Available from: www.espa.ac.uk/files/espa/ESPA-Poverty-Framework.pdf (accessed 13 May 2015).

Folke C, Carpenter SR, Walker B, Scheffer M, Chapin T and Rockström J (2010) Resilience thinking: Integrating resilience, adaptability and transformability. *Ecology and Society*, 15(4), 20.

German Advisory Council on Global Change (n.d.) *World in transition – A social contract for sustainability flagship Report, 2011*. Available from: www.wbgu.de/en/flagship-reports/fr-2011-a-social-contract (accessed 13 May 2015).

Green M (2009) The social distribution of sanctioned harm. In: Addison T, Hulme D, and Kanbur R (eds), *Poverty dynamics: Interdisciplinary perspectives*, Oxford: Oxford University Press, 309–27.

Gunderson LH and Holling CS (eds) (2002) *Panarchy: Understanding transformations in human and natural systems*. Washington, Covelo, London: Island Press.

Hackmann H and St Clair A (2012) *Transformative cornerstones of social science research for global change*. International Social Science Council. Available from: www.worldsocialscience.org/documents/transformative-cornerstones.pdf (accessed 2 December 2014).

Haxeltine A and Seyfang G (2009) *Transitions for the people: Theory and practice of 'Transition' and 'Resilience' in the UK's transition movement*. Tyndall Centre Working Paper. Available from: www.tyndall.ac.uk/sites/default/files/twp134.pdf.

Kates R, Travis W and Wilbanks TJ (2012) Transformational adaptation when incremental adaptations to climate change are insufficient. *Proceedings of the National Academy of Science*, 109(19), 7156–61.

Krishna A (2009) Subjective assessments, participatory methods and poverty dynamics: The stages of progress methods. In: Addison T, Hulme D and Kanbur R (eds), *Poverty Dynamics: Interdisciplinary Perspectives*, Oxford: Oxford University Press, 183–201.

Leach M, Scoones I and Stirling A (2010) *Dynamic sustainabilities: Technology, environment, social justice*. London: Earthscan.

Loorbach D (2010) Transition management for sustainable development: A prescriptive, complexity-based governance framework. *Governance*, 23(1), 161–83.

Marshall N, Park S, Adger WN, Brown K and Howden SM (2012) Transformational capacity and the influence of place and identity. *Environmental Research Letters*, 7(3), 034022.

Maru YT, Fletcher CS and Chewings VH (2012) A synthesis of current approaches to traps is useful but needs rethinking for indigenous disadvantage and poverty research. *Ecology and Society*, 17(2), 7.

Moore M, Tjornbo O and Enfors E (2014) Studying the complexity of change: Toward an analytical framework for understanding deliberate social-ecological transformations. *Ecology and Society*, 19(4), 54.

Nelson D, Adger WN and Brown K (2007) Adaptation to environmental change: Contributions of a resilience framework. *Annual Review of Environment and Resources*, 32(1), 395–419.

O'Brien K (2012) Global environmental change II: From adaptation to deliberate transformation. *Progress in Human Geography*, 36(5), 667–76.

O'Neill S and Handmer J (2012) Responding to bushfire risk: The need for transformative adaptation. *Environmental Research Letters*, 7(1), 014018.

Olsson P, Gunderson LH, Carpenter S, Ryan P, Lebel L, Folke C and Holling CS (2006) Shooting the rapids: Navigating transitions to adaptive governance of social ecological systems. *Ecology and Society*, 11(1), 18.

Park SE, Marshall NA, Jakku E, Dowd AM, Howden SM, Mendham, E and Fleming A (2012) Informing adpatation responses to climate change through theories of transformation. *Global Environmental Change*, 22(1), 115–26

Pelling M (2010) *Adaptation to climate change: From resilience to transformation.* Abingdon and New York: Routledge.

Pogge T and Sengupta M (2015) The sustainable development goals: A plan for building a better world? *Journal of Global Ethics*, 11(1), 1–9.

Ramalingam B (2013) *Aid on the edge of chaos: Rethinking international cooperation in a complex world.* Oxford: Oxford University Press.

Ravallion M, Chen S and Sangraula P (2009) Dollar a day revisited. *The World Bank Economic Review*, 23(2), 163–184.

Scheffer M and Westley F (2007) The evolutionary basis of rigidity: Locks in cells, minds, and society. *Ecology and Society*, 12(2), 36.

Schoon M, Fabricius C, Anderies JM, Nelson M (2011) Synthesis: Vulnerability, traps, and transformations: Long-term perspectives from archaeology. *Ecology and Society*, 16(2), 24.

Steneck R, Hughes T, Cinner J, Adger WN, Arnold SN, Berkes F, Boudreau SA, Brown K, Folke C, Gunderson L, Olsson P, Scheffer M, Stephenson E, Walker B, Wilson J and Worm B (2011) Creation of a gilded trap by the high economic value of the Maine lobster fishery. *Conservation Biology*, 25(5), 904–12.

Tiffen M, Mortimore M and Gichuki F (1994) *More people, less erosion: Environmental recovery in Kenya.* Nairobi: John Wiley & Sons Ltd.

Walker B, Holling CS, Carpenter SR and Kinzig A (2004) Resilience, adaptability and transformability in social-ecological systems. *Ecology and Society*, 9(2), 5.

Westley F, Olsson P and Folke C (2011) Tipping toward sustainability: Emerging pathways of transformation. *Ambio*, 4(7), 762–80.

Wilson S, Pearson L and Kashima Y (2013) Separating adaptive maintenance (resilience) and transformative capacity of social ecological systems. *Ecology and Society*, 18(1), 22.

Woolcock M (2009) Toward an economic sociology of chronic poverty: Enhancing the rigor and relevance of social theory. In: Addison T, Hulme D and Kanbur R (eds), *Poverty dynamics: Interdisciplinary perspectives*, Oxford: Oxford University Press, 328–48.

World Bank (2005) *World Development Report 2006: Equity and development.* Available from: http://bit.ly/1EA2je2 (accessed 24 November 2014).

7   Re-visioning resilience
Resistance, rootedness and resourcefulness

The preceding chapter discussed how resilience concepts can inform understandings of poverty and its persistence, and how international development might be re-shaped in an era of global change. This chapter further develops my vision for resilience informed by diverse science and knowledge, and by the experiences of everyday forms of resilience. Three elements underpin this re-visioning of resilience, and they resonate with and reflect the political ecology perspective laid out in earlier chapters. These three elements are resistance, rootedness and resourcefulness, and they bring new dimensions to current approaches to resilience, which ensure that the concept and the science of resilience speaks more directly and usefully to the concerns of international development and global change. This re-orientation results in a more socially informed understanding of resilience, which acknowledges multiple meanings and understandings, as well as the multi-layered politics and process of dealing with, negotiating and actively shaping change.

Notions of resilience and resistance are related, although usually understood as quite distinct, often antonymic. In practice they are interwoven, and a political ecology approach understands resilience and resistance as potentially allied. Considering how resistance may be an element of resilience brings politics to the core of resilience.

I explore the concept of rootedness in relation to resilience and its relevance to current debates on global change. Rootedness is about place, identity, and belonging, and it might act in different ways to both constrain and support capacities to adapt and transform. I discuss the ways in which factors of place and identity in particular are important in fostering resilience and managing and directing change.

Resourcefulness is not only about the need for people to have access to resources to draw on, but also their capacity to utilise them in the right place and at the right time. Thus, it questions how everyday forms of resilience can be actuated or made to work with different types of change, and their implications of how decisions are made.

Finally, this chapter examines how society might initiate more substantial and fundamental, transformative changes to face the future. It considers how resilience – re-visioned in the ways I suggest – can be central to the negotiation, deliberation, shaping and implementation of the substantive and profound changes necessary for sustainable development.

To begin, I use two vignettes about the experiences of Hurricane Katrina that struck the US Gulf Coast in late August 2005. Much has been written about Hurricane Katrina and the responses to the disastrous flood that it brought to New Orleans and the surrounding areas. These cases are used to set the scene for discussion of this re-visioning and to illustrate the importance of these three elements, which bring to the fore the capacities of people and social ecological systems, and how they respond to many different changes of circumstance. This builds directly on the everyday forms of resilience explored earlier.

New Orleans: a tale of two cities

I visited New Orleans and southern Louisiana in April 2014 to make a short film with National Geographic and AXA Research Fund.[1] Speaking directly to people of their experiences of Hurricane Katrina in August and September 2005, and hearing them recount their experiences nine years after the event, was thought-provoking, inspiring and very moving. People talked a lot about their own, and their community's resilience; about their love for their place; and about the unique culture and society in New Orleans. Each person I spoke to had dramatic and harrowing stories about the hurricane, the floods and the recovery. Most people I met had lost their homes and their possessions, and most had been displaced for months after the event itself. The struggles around rebuilding and recovery were significant – recalling conflicts over access to federal funds and basic infrastructure and other types of support. But I was keenly aware that the people I spoke to were those who had returned and who felt they had a story they wanted to tell, and who wanted to share that story with a visitor from outside their community. But these people had experienced extreme hardship, and some were still very traumatised. As I have reflected on these narratives and experiences, and read more about the Hurricane Katrina phenomenon, it has highlighted some of the key aspects of resilience and its construction by different actors that influence my interpretation, and my insights on the place of resilience and its potential for transformation. So I start this final chapter with two starkly contrasting – but closely related – perspectives on resilience from post-Hurricane Katrina New Orleans.

New Orleans has often been portrayed as a divided, segregated city – a 'paradox'. Before Hurricane Katrina, New Orleans had become a place sharply divided by race and class, with the second highest rate of African American poverty in the USA, and 37 per cent of the city's African American population living in neighbourhoods of concentrated poverty (Gotham & Campanella 2013).

But New Orleans and the experience of Hurricane Katrina illustrates some of the paradoxes around resilience explored in this book – in particular the interplay between resilience and vulnerability on the one hand, and resilience and transformation on the other. It also highlights the apparent divergence between an objective view of resilience, which seeks to measure and assess resilience in quantitative metrics about concrete outcomes, and more constructivist approaches which use peoples' perceptions and narratives about experiences and processes. Significantly, attempts to assess recovery using the first approach generally highlight the lack of resilience in New Orleans, whereas the second approach emphasises the high levels of resilience.

This chapter uses two vignettes – much as they were employed in Chapter 4 – to highlight these and other themes that are developed in this chapter. The vignettes provide rather contradictory views, but also underscore and inform a revised perspective on resilience and its application to understanding development and global change.

As Gotham and Campanella (2013) observe, the failure of the Army Corps of Engineers' levee system during Hurricane Katrina caused flooding and devastation that sparked the largest population displacement in US history, and radically altered the configuration, organisation and character of New Orleans' urban landscape and neighbourhoods. There was also extensive damage along the Gulf Coast in Mississippi and Alabama with 90,000 square miles designated as federal disaster areas. The scale of the damage inflicted is indicated by the following facts and figures:

- Hurricane Katrina, originated in the Atlantic Ocean south of the Bahamas on 23 August 2005; it gathered force over the Gulf of Mexico and became a Category 5 storm, one of the strongest ever recorded in the region. It made landfall in Louisiana on 29 August bringing winds of nearly 200km/hr, 5–10 inches of rain, and a massive storm surge.
- In New Orleans levee breaches caused flooding to over 80 per cent of the urban footprint of the city. It took many weeks for some of these areas to be 'de-watered'.
- Nearly 228,000 housing units – 45 per cent of the metropolitan total – were flooded.
- 80 per cent of the population (452,000) were evacuated before the Hurricane.
- More than one million people in the Gulf region were displaced by the storm. At their peak, hurricane relief shelters housed 273,000 people. Later, approximately 114,000 households were housed in Federal Emergency Management Agency (FEMA) trailers.
- By mid-2006 the city remained with only 50 per cent of its pre-Hurricane population, and just 75 per cent by mid-2012.
- At least 1,836 people were (confirmed) killed. Forty per cent of the deaths in Louisiana were caused by drowning, 25 per cent were caused

by injury and trauma and 11 per cent were caused by heart conditions. Nearly half of the fatalities in Louisiana were people over the age of 74.

- The US federal government has spent $120.5 billion on the Gulf region, post-Hurricane Katrina. The majority of that money, $75 billion, went to emergency relief operations.
- The total estimate of damage for Hurricane Katrina is $108 billion, making it the costliest hurricane in US history (more expensive than Hurricane Sandy in 2012).
- According to FEMA and the National Hurricane Center, Hurricane Katrina is the single most catastrophic natural disaster in US history.

The destruction wrought by Hurricane Katrina was unprecedented but not wholly unexpected. It exposed the utterly inept and inappropriate institutional responses, and the extreme inequality of impacts and recovery processes. Importantly, it also shocked many observers around the world, as the aftermath spread over days and weeks, televised across the globe. We bore witness to the appalling management and the delayed recovery and response, which extended and exacerbated terrible human suffering. This brought questions of resilience and recovery, and responsibility, accountability and blame to the fore. The event has been exhaustively studied, dissected and discussed in literature and public fora. There are public documents and accounts, and extensive libraries detailing experiences and narratives. But as Gotham and Campanella (2013) point out, Hurricane Katrina was just one of a series of events; New Orleans represents a city that has been challenged by a series of exogenous shocks occurring over long-term chronic stress. There have been three major shocks in the past decade: Hurricane Katrina, the 'Great Recession' and the Deepwater Horizon oil spill. Gotham and Campanella describe these as taking place against a backdrop of 'slow burn' stresses such as sea level rise and subsistence, and high rates of social inequality, poverty and racial injustice. Coupled with a rich culture and celebrated heritage, the authors describe New Orleans as a city of 'irony, paradox and contradiction' (2013: 299). This paradox is reflected in analyses of the recovery, some emphasising the slow rates of repopulation, the increased housing costs and crime, others the evidence of 'bouncing back' in terms of economic indicators, employment and businesses starting up. There are thus significant lessons for resilience from these studies and from the mixed experiences and the responses to Hurricane Katrina.

'There's no place like New Orleans': place, culture, community

The Mercatus Center at George Mason University in USA has extensively documented recovery post-Hurricane Katrina in their 'Gulf Coast Recovery Project'. Researchers there have examined the role of sense of place, entrepreneurship and learning in the recovery process. The narrative that emerges from their, and much other, work is one of highly vulnerable poor people impacted upon disproportionately by the Hurricane, and by communities brought together in times of adversity. The quote above forms the title of a paper by researchers on this project, Emily Chamlee-Wright and Virgil Storr (2009). They sought to establish why people returned following evacuation, and the role of place attachment and dependence in motivating people to return and rebuild their communities. Their findings are astonishing, partly because they focused on the poorest areas, where damage was greatest and which received least help. They wanted to know whether sense of place was a critical factor in explaining recovery efforts of residents of the Ninth Ward neighbourhood who faced many structural, systemic and personal obstacles to recovery. Their survey found that more than 84 per cent of respondents attributed their decision to return to a bundle of characteristics that make New Orleans unique in their estimation. The distinct sense of place which residents articulated was not related to any one attribute, but people mentioned a range of elements, including social networks based on friends, neighbours and church congregations; around food and music; jazz funerals and second-line parades; and a climate which meant people spend much time outdoors, and develop social interactions with neighbours and passers-by from their front porches. This is very similar to the accounts people gave me, recounting stories about neighbourhood barbeques and great feasts of shrimp, crawfish or crab. People rejoiced in the apparent spontaneity and sharing of these impromptu gatherings.

When evacuees were asked why they had returned, people explained that their quality of life would be poorer without these unique aspects of New Orleans. But they often had to weigh this against the lack of services and infrastructure (particularly schools) and the economic costs of returning to a city that remained poorly provisioned for years. Chamlee-Wright and Storr remark that for many people who didn't return to their neighbourhoods, it was because their sense of responsibility to family and need for education and health care over-rode the pull of the uniqueness and place attachment to the city. This signals the need to re-open schools and hospitals quickly, but also demonstrates the interplay of the community and state and the personal and the public in determining peoples actions and ultimately the trajectory of their recovery.

Chamlee-Wright and Storr argue that in times of contingency, sense of place becomes transformed by the disaster from background context to an important cultural resource. Indeed this was strongly articulated by some of the people we spoke to when we were making our film, shown in Box 7.1. The experience of Hurricane Katrina, one of the many storms, shocks and destabilising forces faced by people in southern Louisiana and New Orleans, added to peoples' sense of self-efficacy, their agency and their belief in their capacity to deal with other shocks. Their experience affected their perception of community resilience. Lucy Faulkner (2014) found a similar phenomenon

---

Box 7.1 Expressions of place attachment and community resilience in New Orleans

'If another Katrina came again, you would really see how strong and how together we would be.'

'When people come together with the same purpose and have the same vision, there's nothing that can't be done.'

'If another catastrophe happened, worse than Katrina, the only thing I'm gonna do is leave. But then I'm gonna come back and rebuild my house. Because I love my area. It's worth saving.'

Source: National Geographic/AXA 2014

---

when she interviewed people in two communities in north Cornwall in the UK. Where people had experienced a flood within the last decade, they had much more confidence about their ability and capacities in their community to deal with a range of changes – they assessed community resilience as at a higher level than in a neighbouring site which had not been recently flooded. This has been observed in other studies, and is sometimes referred to as an 'inoculation effect'.

Rebecca Solnit (2010) catalogues these reservoirs of community strengths and resilience in her remarkable book, *A paradise built in hell*. Solnit describes what happened after the Hurricane, after the levees broke and after the failed relief efforts, where the poor and vulnerable were abandoned, as a 'socio-political catastrophe' and a huge crime and national shame (p141). Whilst recounting the violence and the failures of resource and relief efforts, her book is actually about the way in which people helped each other, how citizens and communities – completely contrary to media reports at the time – displayed astonishing bravery, altruism, generosity and resourcefulness. She documents 'the extraordinary communities that arise in disaster', which surely point to the opportunities for transformative change, for utilising hitherto latent capacities beyond the immediate aftermath and context of the extreme event, the calamity, the shock or disaster. Gotham and Campanella (2013) discuss the extent to which the recovery of New Orleans post-Hurricane Katrina can be seen as transformative, and comment that resilience is 'not a static descriptor of a neighbourhood but is expressed in actions', resting often on the capacities of individuals working together. These observations all point to the potential transformative capacity of collective resilience.

'Stop calling me resilient': resistance, power and accountability

Figure 7.1 'Stop calling me resilient': New Orleans fly-posting

Source: Photographer unknown

This image shows a notice posted in New Orleans.[2] It reads: 'Stop calling me resilient. Because every time you say, "Oh, they're resilient" that means you can do something else to me. I am not resilient.'

This notice articulates a core criticism of resilience as hijacked by a neoliberal political agenda, which shifts the burden of recovery – and implied blame – to individuals and communities themselves. The logic, clearly spelled out in the notice, is that if communities are resilient, then they can absorb shocks and do not need protection.

Much of the writing around Hurricane Katrina has highlighted the inequalities and injustices associated with its impacts, recovery and rebuilding. New Orleans was a highly unequal city before the event itself, stratified and fractured along economic, racial and class lines, played out in the geography of its neighbourhoods. Yet even in 2010, poverty rates for African Americans were more than double that of white Americans, but nearly seven times that for children under 17 years of age. Black residents were geographically isolated in poor neighbourhoods which experienced greater flooding and population loss after the flood. So poor people were concentrated in areas which were most flooded and lived in generally poorer-quality housing. They were clearly more vulnerable in these respects. But also poor people, and black people, were further disadvantaged in the aftermath and in the rescue and recovery phases.

A number of studies of these processes have focused on the Lower Ninth Ward, a poor neighbourhood which was badly flooded, with up to ten feet of flood water, amongst the worst in the city. By 2010 the area had recovered only 20 per cent of its pre-Hurricane Katrina population, with only 2,842 residents compared with more than 14,000 before the flood (Gotham & Campanella, 2013). During 2007 many of the houses were bulldozed, producing some of the abandoned and razed areas I was shocked to see when I visited there in 2014.

Gotham and Campanella's (2011) analysis of cross-scale pathologies identifies some of the structural forces at work which reinforced inequalities and created constraints and obstacles to recovery and rebuilding. Whilst numerous and diverse institutions – state, federal, non-profit, private businesses and foundations – worked together in these processes, the linkages and interactions between them were vital in determining who and where got support. The authors used resilience concepts, including connectivity and cross-scale linkages, to examine the extent to which patterns of vulnerability and resilience were changed, and whether Hurricane Katrina provided a 'window of opportunity' to change existing social and ecological systems. Overall, they conclude that the cross-scale interactions between the institutions and agencies produced negative feedbacks and maladaptations which perpetuated social inequalities and reinforced pre-existing vulnerabilities. In addition to the obstacles to funding (which include lack of access to information, lack of trust and accountability) the analysis reveals how contested property rights, land claims and land use planning processes both complicate and problematise the adaptive capacity of communities. These factors combined ultimately jeopardise longer-term social and ecological sustainability.

In his book *Floodlines: Community and resistance from Katrina to the Jena Six*, Jordan Flaherty (2010) exposes the overt racism and social injustices of the organised and state responses in the aftermath of Hurricane Katrina in New Orleans. He chronicles the systematic and institutionalised racism and violence pitted against flood victims as they sought shelter and as they tried to leave the city. He witnessed the plight of people stranded on rooftops and bridges, and he spent time at a makeshift camp near the highway, Interstate 10, where 95 per cent of the evacuees were African American and everyone waited under armed guard behind metal barricades. Importantly he documents how the rebuilding of the city has eroded rights and has deepened inequalities; he highlights the erosion of the public health care, criminal justice, housing and education system. But he also examines how this has inspired and prompted a coalition of grassroots activists, direct action, lobbyists and civil rights lawyers to form a social movement around issues of justice and resistance. In galvanising this movement and in shining a spotlight internationally on New Orleans, Hurricane Katrina has unleashed and ignited community activism which inspires perhaps a social transformation much as the resilience and poverty literatures present in theoretical terms, putting resistance at the heart of these social changes. This is expressed by Flaherty (2010: 38) in the following:

Those who have not lived in New Orleans have missed an incredible, glorious, vital city – a place with an energy unlike anywhere else in the world, a majority-African American city where resistance to white supremacy has cultivated and supported a generous, subversive, and unique culture of vivid beauty. From jazz, blues, and hip-hop to second lines, Mardi Gras Indians, jazz funerals, and the citywide tradition of red beans and rice on Monday nights, New Orleans is a place of art and music and food and traditions and sexuality and liberation.

These two vignettes give quite different perspectives on resilience of New Orleans post-Hurricane Katrina. The first emphasises individual agency and collective action, and the importance of place and identity. The second shows the structural reasons why recovery was so difficult and why the impacts exacerbated inequalities, and the systemic causes of vulnerability. They show, vitally, that we need an approach to resilience which combines a concern for agency and individuals and collective action, with an understanding of systemic dimensions of broader drivers and mediators of change. This echoes the discussion of political ecology started in Chapter 1. This implies that the everyday forms of resilience are important, but they must be accompanied and supported by institutional and higher order support. But the vignettes also reinforce the dynamic interconnections between people and place, and society and environment, that in part determines the capacity, or agency, to respond to and to shape change.

Re-visioning resilience

We study the extraordinary to understand the ordinary, and we examine the ordinary to inform the extraordinary. I use these vignettes about responses to Hurricane Katrina and its aftermath, most extraordinary events, to highlight some of the diverse facets of resilience. Just as the scientific and policy literatures discussed in earlier chapters points out the mixed, diverse and conflicting conceptions and indicators of resilience, so the experiences and perceptions of New Orleans residents demonstrate how they construct a wide range of meanings and narratives around their recovery. Here I show how a more socially informed and pluralist perspective speaks to a political ecology of resilience. This highlights the politics of how people can respond, how the capacity of social ecological systems depends on a wide range of different attributes, how institutions and social norms operate, the politics of who gets what and the sociology, economics and psychology of who stays, who goes, and when or if they return. It highlights how crudely, but also how subtly and how dynamically, there are 'winners and losers'.

Here I re-vision resilience for development and put an emphasis on everyday forms of resilience. This builds directly on the lived experiences and the narratives introduced in the vignettes in Chapter 4. This is applied, as Amin (2013: 141) suggests, to de-dramatise and de-centre the neoliberal prospectus, and put emphasis on human agency, collective action and knowledge, and 'everyday know-how', constituting the 'contingent vitality' that Amin talks of in relation to urban settlements. This re-vision addresses the criticisms of the resilience approach in social ecological systems as over-emphasising the structural and external forces, and negating the agency of individuals and households, and their different aspirations, preferences and power relations.

A focus on everyday forms of resilience gives a starting point to truly integrate understandings of how people can respond to change – both the slow variables, such as changes in values, and the shocks. It puts the capacities that people, communities, social ecological systems have at the core of the analysis. It follows on from Rebecca Solnit (2010), who identifies social inventiveness, altruism and solidarity as being key characteristics of peoples' responses to disasters. The three issues that are core to this re-visioning of resilience are resistance, rootedness and resourcefulness. These form the bedrock of what makes people, their communities and social ecological systems able to respond effectively, and often positively, to change.

Resistance

Strongly resonating with the earlier discussions about political ecology and reflecting on the narratives of resilience discussed in Chapter 4 this section suggests that resistance may be an important element of resilience. This is particularly the case if we view resistance as a potential site for change and the means through which individuals change social processes and structures and build alternatives.

It is striking that often in my research when we speak to individuals or hold focus groups in communities about resilience, people refer to resilience as a form of resistance. Whilst in its simplest form this might imply the ability to resist change, in most cases resistance refers to the ability or capacity of people to withstand external forces and to actually shape their own strategies. Resistance here implies strength, self-determination, agency and power. As discussed in Chapter 4, the narratives of people in the Orkney islands emphasised their capacity to withstand outside interventions, to deal with the weather and the sea, and their desire to nurture self-sufficiency and local democracy. In Mozambique people expressed their vulnerability in terms of their inability to challenge government policies and the internationally-initiated and led projects and developments that affected their livelihoods. This establishes a direct link between resilience, agency and power, and resistance. Resilience as resistance also encompasses the power to resist change determined by outsiders and imposed by outside forces; relating resilience to empowerment and self-determination.

Resistance is about the power or capacity to resist. But resist what? It means very different things in natural and social sciences. In ecology resistance is the ability of communities or populations to remain essentially unchanged in the face of disturbance. It is closely allied to the concept of resilience, although it is associated with stability rather than adaptability. But in social sciences – primarily in sociology, political science and geography – resistance is generally understood as concerned with undermining power relations, or about the creation of or expansion of space for decision-making. It is often understood to be oppositional – about the exploited against the exploiter, the dominated against the dominator, or the oppressed against the oppressor – it is about the exercise of subordinate power. But resistance might also involve reforming cultural norms or challenging conventional values. Resistance is often associated with political or social movements, as organised efforts by portions of civil society to resist and to disrupt order and stability. James Scott (1987) highlighted that most resistance is hidden, informal and non-confrontational.

There exists a dialectic relationship between resilience and resistance; for example, they have distinct meanings in ecology that cannot be transferred to the social sphere or even social ecological domain. Applied in disasters studies, resistance is often seen as complementary to resilience – where resistance relates to the capacity to resist the onset and minimise the impacts of disasters, whereas resilience emphasises the capacity to absorb and recover from the effects. Thus resistance would involve building levees to control floods, and civil engineering projects such as sea walls to prevent coastal inundation.

Emphasising the complementarity of resilience and resistance, and understanding resistance as an element of resilience, also propels us to a more proactive and potentially transformative view of resilience. For example, Bottrell (2009: 337) notes that 'including resistances in the conceptualization of resilience suggests the need for change in positioned perspectives, structured inequalities and the distribution of resources for strengthening resilience', in other words, transformative change. In conceptualising communities as agents of change, Chaskin (2008) presents community resilience in three forms – one of which is resistance (the other two being re-grouping and redevelopment) reflecting a community's ability to take action together and to shape the factors affecting them, including external policies.

As Beymer-Farris et al. (2012) emphasise, power relations and ecological conditions structure social ecological systems and produce unequal outcomes that are often highly contested by different – and competing – resource users. Their approach – as mine – seeks to integrate a political ecology, or political economy, analysis with resilience insights to better understand these dynamics. Their examination of the expansion of prawn farming in Mafia Island in Tanzania shows an important role of resistance; in how local fishers are able to challenge the activities of industrial prawn farmers, and in how a 'desirable' state is defined and shaped. They meticulously map these processes and the contestations around social ecological change to the adaptive cycle. They document the rise of prawn farming during the expansion phase, then more widespread and damaging mangrove clearance and increasing pollution that gradually lead to a maladaptive shift, so that the prawn fishery becomes stuck in a rigidity trap, and then the emergence of a social movement to challenge the dominance of prawn farms. They argue that power relations, particularly those supporting economic growth and influencing access, control and management of resources are key to defining the desirable state. So resistance informs both the capacity and strategies of actors to influence slow and fast variables at different points in the adaptive cycle. This shows that the approaches are complementary and also that resistance is articulated through the actions and responses of resource users in different ways.

Considering resistance as an element of resilience necessitates that power be explicitly examined. It shifts emphasis towards the drivers or causes of change, and the structural determinants of vulnerability and resilience. Each of these factors has been highlighted as missing from the social ecological systems literature; yet they feature in the community development and human development literature on resilience. Because resistance is about capacity and strength it resonates with human development and the 'strength-based' approaches rather than deficit-based models. But it also recognises the fundamental role of contestations – about access to resources, a desired state, about possible futures, and intra- and inter-generational equity – in negotiating change in social ecological systems. Above all it puts agency at the heart of resilience.

Rootedness

In New Orleans people affirmed 'There's no place like New Orleans' but the place they referred to went way beyond a physical site. It included the social networks, the cultural practices, and a wide range of affective ties with 'home'. Christopher Lyon (2014) develops a framework for understanding place in social adaptation, resilience and transformation. He disaggregates 'place' to analyse how its different dimensions relate to resilience. Most studies, he observes, focus on sense of place or place attachment. He defines three dimensions of place: incarnate place or physical place; discarnate place or place character, including heritage and cultural aspects of place which enable or constrain local agency; and chimerical place, or place attachment and sense of place. As discussed in Chapter 6, a number of studies have shown attachment to place, and place-rooted identity to be powerful determinants of adaptation, resilience and transformation (Devine-Wright 2013). But material aspects of place – the natural and built environment and associated infrastructure and services – have been shown to be important for resilience and adaptive capacity (Chapter 5, and e.g. Berkes & Ross 2013). In addition the importance of cultural dimensions – discarnate place – has been emphasised in adaptation (Chapter 6, and Adger et al. 2012). My conceptualisation of rootedness encompasses each of these dimensions.

Amin (2013: 141) talks about 'situated resilience', asserting that the 'turbulent future will be addressed through the specifics of location'. Rootedness reflects this, and emphasises the power of place, community and identity in resilience. It reflects the strengths associated with belonging. When Lucy Faulkner (2014) interviewed people in Cornwall in the UK about their perceptions of community resilience, her analysis revealed that sense of place and place attachment were right at the heart and ranked as the most important pre-condition for building resilience.

Rootedness is more than place, partly because people are increasingly mobile, so it is about a more fluid set of attachments and multi-faceted identities. Robin Broad and John Cavanagh (2011) too view rootedness as encompassing a much broader set of aspects at multiple scales, and go as far as to suggest rootedness as a new paradigm for development in the age of rapid change and recurrent crisis. Three recent and recurring crises – of food, finance and environment – expose the weakness and fault-lines in the conventional development model. They posit rootedness as an alternative to vulnerability, as a strength-based approach to understanding how people, communities and economies might thrive. They reject resilience as an organising principle – on the grounds that it is too narrow in its application, and focuses too much on bouncing back after shocks. They see rootedness as encompassing contemporary development concerns for human rights, ecological sustainability, participatory democracy and redistribution and equity. Economic, environmental and social rootedness engenders sustainability, sustainable livelihoods, resilience and subsidiarity. They suggest thirteen indicators that can assess a continuum between vulnerability and rootedness. All households, communities and countries fall along this spectrum, and human well-being is expected to be higher in those households, communities or countries with higher rootedness. In their view rootedness encompasses (Broad & Cavanagh 2011):

- *Economic rootedness*, a form of subsidiarity which focuses on producing as much as possible locally, then nationally, then regionally, and only then globally. It relates to localism and security in production, self-sufficiency and vibrant local economies.
- *Environmental rootedness* puts community control over natural resources, such as water and forests, at its core, providing incentives for sustainable management.
- *Social rootedness* emphasises equity and health in order to foster a sense of community.

The dictionary definition of rootedness relates to the quality or state of being well-grounded. Rootedness refers to intertwined people and place, or a sense of belonging somewhere. In psychology, rootedness represents the strongest bond between people and their community or place. But, like resilience itself, rootedness is not always positive. It might be cohesive or divisive – it might bring people together and form strong bonds, or it may divide people. In psychology, rootedness is the need to establish roots and to feel at home in the world; to establish ties with the outside world. It is therefore also about our connection with the ecology of place, and our sensitivity to changes in social ecological systems, and our reactions to this. In this way, groundedness does not equate to stasis or stability, it also means responsiveness and adaptiveness.

Resourcefulness

Resourcefulness encompasses the resources that people can draw on, but also the capacity to use them at the right time, in the right way. As the ability to meet and deal with difficult situations inventively, it encompasses human agency and capabilities, and it is about opportunities and about ideas, and innovation. Thus, for Gotham and Campanella (2013: 300) resilience in post-disaster situations such as New Orleans 'lies in the ability of groups, organizations and institutions to access and effectively use political, economic, cultural and natural resources for recovery transformation and innovations'. Ealy (2009) documents the resourcefulness of community-based organisations in responding to Hurricane Katrina, and talks of the nimbleness of community and faith-based organisations. They were able to be responsive and build on local knowledge and social networks, and improvisational ability that no large, complex outside agency – federal or state – could muster. They were also able to mobilise and co-ordinate many individual acts of spontaneous action, mobilising the personal and collective efforts, matching Norris et al.'s (2008) definition of community resilience as a 'process linking a set of adaptive capacities' (see Figure 3.4). Resourcefulness is about linking those pre-existing assets and capacities within a community or associated with a place, to external streams of resources, and making sure that these are used effectively. This chimes with Chamlee-Wright and Storr's (2009) earlier observation of sense of place being transformed into a resource in times of need. Resourcefulness is about bouncing back, adapting and transforming.

Resourcefulness in this respect links to innovation, social learning and social capital, all of which have been identified as key components of adaptive capacity and resilience. In encompassing this range of attributes, resourcefulness implies assets, capacity and elements of timeliness and initiative. In Kenya our research with coastal communities has asked people about their capacity to deal with change, and what they need to 'live well'. Many people – both men and women – talk about the 'developmental mind' (*mawazo* in Swahili) which encompasses a similar notion. It means a person needs to be able to access resources, to have the confidence and capacity to use them, and needs good social and business networks, and good relations with family, neighbours and workmates. It is partly an entrepreneurial spirit, partly good local knowledge and business acumen. But it relies on social ties too. As a fisher told us in a focus group on well-being: 'Getting a job depends on luck – but you may have luck, yet if you don't have *mawazo* then you will not progress – this is because there is a lot of choices and it is important to make the right choice and invest on business that can generate money'.

Resourcefulness means that resources and capacity must be harnessed together. Again it is an element of resilience which links the social and ecological parts of the social ecological system, and like adaptive capacity it might be drawn on in times of need, such as in sudden events, shocks or surprises, but is also employed to strategise, plan and manage change on a daily basis. It is part of everyday resilience as well as necessary for times of adversity. Resourcefulness also relates strongly to what Rival (2009) referred to as 'indigenous intelligence', (see also Chapter 4) and puts the ingenuity and practices associated with local knowledge and lay knowledge into context.

Re-conceptualising a social ecological system

Much of the discussion in this chapter has presented a very human-centred view of resilience. But what then of the ecosystem itself, and of the social ecological system? Most representations of a social ecological system present two sub-systems – the social and the ecological – interacting within a larger arena, the social ecological system. Various linkages, interactions and feedbacks between the two sub-systems are posited. These are mediated by, for example, institutions such as property rights that govern people's access to and control over different components of the system. In many figures and diagrams in the literature these are denoted as one-way and two-way arrows between the two sub-systems. But what characteristics are shared across the two sub-systems and how can we understand resilience better as a property of the social ecological system rather than of the two sub-systems? It is striking that when we have developed mental models and representations of social ecological systems with stakeholders, for example, in the Tanzanian and Mozambique examples discussed in Chapter 4, and in recent work in Kenya (Daw et al. 2015), people visualise the system holistically, and not divided between the ecological and the social.

My re-visioning of resilience emphasises agency, but chiefly agency of human actors in the social ecological system. But this idea of agency might be extended beyond humans. In a recent paper, Dwiartama and Rosin (2014) propose that actor network theory might provide a useful starting point to extend agency to non-humans to develop a more tightly-coupled view of a social ecological system. Indeed in an earlier discussion, Murdoch (1998) proposes actor network theory as a means of navigating dualisms such as nature/society, action/structure and local/global, which implies that the approach can inform social ecological systems. Murdoch highlights the heterogeneity of networks, and their concern with how social and material processes are intertwined within complex sets of association – what he refers to as network topologies. Christmann et al. (2012) also consider actor network theory in relation to vulnerability and resilience. They view that emphasising agency not just of individual actions, but of associations and networks as dispersed competencies, can inform and overcome social ecological dichotomies within the social ecological system concept.

Dwiartama and Rosin (2014) argue that actor network theory can inform resilience analysis, by offering the opportunity of a more encompassing view of agency that extends beyond human intentionality. This focuses on the relationships in which agents participate and how these influence the shape of a network of relationships. In actor network theory agency can be extended to non-humans, including animals, materials, ideas and concepts. Thus diverse components of a social ecological system, including plants and animals, minerals and climate, are system-forming entities. This enables perhaps the role of relations between humans (the social sub-system) and non-humans (the ecological sub-systems) in resilience dynamics to be viewed holistically and as an emergent property of the larger social ecological system itself.

Whilst not promoting actor network theory in particular, developing an extended conceptualisation of agency to integrate the social and the ecological and to understand the form of linkages and influences would be a step forward for social ecological systems resilience, and enable a more dynamic view of the social as well as the ecological (recalling the comments of Hatt (2013) and others discussed in Chapter 1). Extending agency – and re-examining the ways in which different components of the social ecological system perform and have influence on outcomes – can be applied to the examples outlined in the vignettes from Chapter 4. In Cameroon the market is a key actor shaping the responses of women and men, and determining the outcomes for different people, places, and the ecology of the forest. In northern Tanzania, the marine protected area influences how human actors and different habitats interact. In Laura Rival's (2009) concept of indigenous intelligence, the knowledge networks of different ethnic groups are key for social and ecological outcomes.

In Brown and Westaway (2011) we develop Lister's (2004) taxonomy of agency to understanding responses to climate change. This identifies that positive transformative responses require moving from an individual to a collective, and an everyday to a strategic form of agency. This might imply that the extension of agency beyond human actors gives greater insights into the possibilities for transformation of the social ecological system.

The three elements suggested here – resistance, rootedness and resourcefulness – are not just concerned with properties of the social or human dimensions of a social ecological system. Resistance is a process linking the social and the ecological as Beymer-Farris et al.'s (2012) re-worked adaptive cycle demonstrates, and is also a property of non-human, material components. Rootedness, firmly associates people and place or space. Resourcefulness concerns the material aspects of the social ecological system – often of the ecosystem itself – as well as human capacity and ingenuity to manage them.

Folke et al. (2011: 719) make an emphatic case for 'reconnecting with the biosphere' to realign human development with the global social ecological system. Whilst clearly referencing a 'planetary boundaries' framing for resilience, they recognise that reconnections are not just material, organisational and physical. Reconnecting with the biosphere demands a re-shaping of relationships based on a shift in worldviews and values, attitudes and emotions. Although they focus on the concept of planetary stewardship, a more normative suggestion for reconnections a re-conceptualisation of a fully integrated social ecological system provides scope and opportunity for reconnecting humans and nature at multiple scales.

Deliberating the future

How does a resilience perspective help us to make decisions and face the future? If we re-vision resilience and take an agency-orientated perspective, then it provides an opportunity to reconsider the process of deliberation and the negotiations about desired futures and actions. I've shown how resilience is not value-free but has multiple meanings and contestations and this has important implications for how resilience might be applied and used in decision-making and governance. This re-visioned perspective puts emphasis on the process of change, on how decisions are made and actions taken, in contrast to much resilience work which emphasises the outcome of actions (Robards et al. 2011). In setting a scientific agenda to inform the transformations necessary to address global environmental change, Karen O'Brien (2012) asks how transformations can be carried out in a deliberative, participatory manner that is both ethical and sustainable. What are the opportunities for a resilience approach to contribute to this process and to negotiating the new social contract demanded by climate and other global change (O'Brien et al. 2009)?

A focus on process leads back to a consideration of the relationships between the social and ecological and between different scales, including spatial and temporal. Bruce Goldstein et al. (2013) develop a notion of collaborative resilience and have used resilience thinking in collaborative planning in urban contexts and with forest and natural resource managers to control fires. This approach recognises that conventional resilience analysis does not engage with the material, social and symbolic landscape that constitutes the lived experience of the communities whose resilience is being sought. Hence the discussions here of everyday forms of resilience arising from the analysis of the lived experiences of resilience in Chapter 4, which has been central to my re-visioning. Goldstein et al. too seek to completely overturn this omission, using collaborative planning and narratives analysis and a range of participatory methods. Thus, they claim that 'resilience can serve as a conceptual framework for exploring and enabling new urban possibilities, moving beyond the goals of recovery or persistence that characterise much of sustainability thinking' (2013: 2). For Goldstein and colleagues, 'Resilience thinking highlights the futility of predictive forecasts based on assumptions of order, certainty and equilibrium, an "engineering" mode of operation that is still deeply embedded in planning method and practice' (p2), and instead offers language, ideas and methods that account for the dynamic character of complex systems. This necessitates methods that engage multiple voices and accommodate multiple views. It suggests that adaptive capacity is built through collaborative problem solving, social learning and engaging a diversity of stakeholders and knowledge practices. But these processes are also intensely political themselves, as are the tensions between scientific and other knowledges and inherent power asymmetries (Robards et al. 2011). Yet a process-orientated view of resilience must explicitly address these issues, as well as the social construction of resilience. So resilience can become a way of opening up new opportunities which must be considered and discussed and deliberated upon by various stakeholders – those who are affected and those that influence decisions. We have used a range of different techniques, including developing scenarios and participatory modelling, focus groups and different deliberative fora about sustainable responses to change, and to identify winners and losers and trade-offs associated with different courses of action (Daw et al. 2015). The point is, resilience should open up, not close down, space for negotiation.

Resilience concepts are being used in this way around the world to start processes of transformative development – for example by the Transition Movement[3] worldwide, by national groups such as SNIFFER,[4] and by a plethora of community groups.[5] The three elements suggested here – resistance, rootedness, resourcefulness – are key to these transformative social movements. They challenge the status quo and the structural determinants of unsustainable energy, social policy and food systems. They promote rootedness through localism and development of thriving local economies, food systems and alternative currencies. They foster resourcefulness in building skills, learning and information networks.

In the re-visioning proposed here, resilience, and its multiple dimensions and properties, becomes a platform to support the deliberation around strategies to negotiate and shape current and future – safe and just – sustainabilities. It contributes substantially and vitally towards – but does not in itself constitute – a theory of change and a new paradigm for development in the era of global uncertainties.

Notes

1   www.youtube.com/watch?v=HxWf9yE6rT4.

2 The quote in the notice is assigned to Tracie L. Washington a lawyer and activist from the Louisiana Justice Institute which has been campaigning for justice for citizens of New Orleans: www.louisianajusticeinstitute.org.
3 www.transition.org.
4 www.sniffer.org.uk.
5 An example is the Carnegie Trust, working with a range of community groups in the UK www.carnegietrust.org.uk.

References

Adger WN, Barnett J, Brown K, Marshall N and O'Brien K (2012) Cultural dimensions of climate change impacts and adaptation. *Nature Climate Change*, 3(2), 112–17.

Amin A (2013) Surviving the turbulent future. *Environment and Planning D: Society and Space*, 31(1), 140–56.

Berkes F and Ross H (2013) Community resilience: Toward an integrated approach. *Society and Natural Resources*, 26(1), 5–20.

Beymer-Farris B, Bassett, T and Bryceson I (2012) Promises and pitfalls of adaptive management in resilience thinking: The lens of political ecology. In: Plieninger T and Bieling C (eds), *Resilience and the cultural landscape: Understanding and managing change in human-shaped environments*, Cambridge: Cambridge University Press, 283–300.

Bottrell D (2009) Understanding 'marginal' perspectives towards a social theory of resilience. *Qualitative Social Work*, 8(3), 321–39.

Broad R and Cavanagh J (2011) Reframing development in the age of vulnerability: From case studies of the Philippines and Trinidad to new measures of rootedness. *Third World Quarterly*, 32(6), 1127–45.

Brown K and Westaway E (2011) Agency, capacity, and resilience to environmental change: Lessons from human development, wellbeing, and disasters. *Annual Review of Environment and Resources*, 36(1), 321–342.

Chamlee-Wright E and Storr V (2009) 'There's no place like New Orleans': Sense of place and community recovery in the Ninth Ward after Hurricane Katrina. *Journal of Urban Affairs*, 31(5), 615–34.

Chaskin RJ (2008) Resilience, community, and resilient communities: Conditioning contexts and collective action. *Child Care in Practice*, 14(1), 65–74.

Christmann G, Ibert O, Kilper H, Moss T (2012) Vulnerability and resilience from a socio-spatial perspective: Towards a theoretical framework. IRS Working Paper 45, Leibniz Institute for Regional Development and Structural Planning. Available from: www.irs-net.de/download/wp_vulnerability.pdf (accessed 13 August 2015).

Daw TM, Coulthard S, Cheung WWL, Brown K, Abunge C, Galafassi D, Peterson GD, McClanahan TR, Omukoto JO and Munyi L (2015) Evaluating taboo trade-offs in ecosystems services and human well-being. *Proceedings of the National Academy of Science*, 112(22), 6949–54. Available from: www.pnas.org/cgi/doi/10.1073/pnas.1414900112 (accessed 13 August 2015).

Devine-Wright P (2013) Think global, act local? The relevance of place attachments and place identities in a climate changed world. *Global Environmental Change*, 23(1), 61–9.

Dwiartama A and Rosin C (2014) Exploring agency beyond humans: The compatibility of Actor-Network Theory (ANT) and resilience thinking. *Ecology and Society*, 19(3), 28.

Ealy LT (2009) Coordinates of resilience: On the nimbleness of community and faith-based organisations in disaster response and recovery. Available from: http://

localknowledge.mercatus.org/articles/coordinates-of-resilience (accessed 19 March 2015).

Faulkner L (2014) Assessing community resilience in north Cornwall. MRes dissertation, University of Exeter.

Flaherty J (2010) *Floodlines: Community and resistance from Katrina to the Jena Six.* Chicago: Haymarket Books.

Folke C, Jansson Å and Rockström J (2011) Reconnecting to the biosphere. *Ambio*, 4(7), 719–38.

Goldstein B, Wessells A, Lejano R and Butler W (2013) Narrating resilience: Transforming urban systems through collaborative storytelling. *Urban Studies*, 52(7), 1285–1303, 0042098013505653.

Gotham K and Campanella R (2013) Constructions of resilience: Ethnoracial diversity, inequality, and post-Katrina recovery, the case of New Orleans. *Social Sciences*, 2(4), 298–317.

Gotham K and Campanella R (2011) Coupled vulnerability and resilience: The dynamics of cross-scale interactions in post-Katrina New Orleans. *Ecology and Society*, 16(3), 12.

Hatt K (2013) Social attractors: A proposal to enhance 'resilience thinking' about the social. *Society and Natural Resources*, 26(1), 30–43.

Lister R (2004) *Poverty*. Cambridge: Polity.

Lyon C (2014) Place systems and social resilience: A framework for understanding place in social adaptation, resilience, and transformation. *Society and Natural Resources*, 27(10), 1009–23.

The Mercatus Center at George Mason University (n.d.) Gulf Coast Recovery Project. Available from: http://mercatus.org/gulf-coast-recovery-project (accessed 18 March 2015).

Murdoch J (1998) The spaces of actor-network theory. *Geoforum*, 29(4), 357–74.

National Geographic/AXA (2014) Explore/protect, K Brown: Community resilience and extreme weather events. Available from: www.katrinabrown.org/2014/09/community_resilienceextreme_weather.

Norris F, Betty SP, Pfefferbaum B, Wyche KF and Pfefferbaum RL (2008) Community resilience as a metaphor, theory, set of capacities, and strategy for disaster readiness. *American Journal of Community Psychology*, 41(1–2), 127–50.

O'Brien K (2012) Global environmental change II: From adaptation to deliberate transformation. *Progress in Human Geography*, 36(5), 667–76.

O'Brien K, Hayward B and Berkes F (2009) Rethinking social contracts: Building resilience in a changing climate. *Ecology and Society*, 14(2), 12.

Rival L (2009) The resilience of indigenous intelligence. In: Hastrup K (ed.), *The question of resilience: Social responses to climate change*, Copenhagen: The Royal Danish Academy of Sciences and Letters, 293–313.

Robards M, Schoon M, Meek C and Engle, NL (2011) The importance of social drivers in the resilient provision of ecosystem services. *Global Environmental Change*, 21(2), 522–29.

Scott J (1987) *Weapons of the weak: Everyday forms of peasant resistance.* New Haven, CT: Yale University Press.

Solnit R (2010) *A paradise built in hell: The extraordinary communities that arise in disaster.* London and New York: Penguin.

Washington TL (n.d.) Louisiana Justice Institute. Available from: www.louisiana justiceinstitute.org (accessed 19 March 2015).

Glossary

This glossary was compiled using the following sources:

Folke C, Carpenter SR, Walker B, Scheffer M, Chapin T and Rockström J (2010)
    Resilience thinking: Integrating resilience, adaptability and transformability.
    Ecology and Society, 15(4), 20.
The Resilience Alliance Glossary: www.resalliance.org/index.php/glossary.
Walker B and Salt D (2006) *Resilience thinking: Sustaining ecosystems and people
    in a changing World*. Washington, Covelo, London: Island Press.

adaptability   The capacity of actors in a system to manage resilience.
adaptive capacity   The capacity to adapt to and shape change.
adaptive cycle   A metaphor or heuristic device which describes how a social
    ecological system changes through various phases of organisation and
    function. It distinguishes four phases: exploitation, conservation,
    creative destruction, and renewal (also referred to as *r*, *K*, *omega*,
    alpha).
adaptive management   A systematic process for continually and iteratively
    adjusting policies and practices by learning from the outcome of
    previously used policies and practices. Each management action is
    viewed as a scientific experiment designed to test hypotheses and probe
    the system as a way of learning about the system.
boundary object   An entity shared by several different communities, but
    viewed or used differently by each of them. Boundary objects are said to
    allow co-ordination without consensus as they can allow an actor's
    local understanding to be re-framed in the context of a wider collective
    activity, and to link communities on a common task.
complex adaptive systems   Systems of people and nature in which
    complexity emerges from a small set of critical processes that create and
    maintain the self-organizing properties of the system.
drivers   External forces or conditions which cause change in a system.
ecosystem services   The benefits that people derive from the ecosystem.
    These include provision, regulating, provisioning and cultural
    services.

engineering resilience   A measure of the rate at which a system approaches steady state following a perturbation, also measured as the inverse of return time.

equilibrium   A steady-state condition of a dynamic system – where variables are interacting but not changing.

experiential   Relating to or resulting from experience.

feedbacks   The secondary effects of a direct effect of one variable on another, causing change in the magnitude of that effect. A positive feedback enhances the effect; a negative feedback dampens it.

general resilience   The resilience of any and all parts of a system to all kinds of shocks, including novel ones.

governance   The structures and processes by which people in societies make decisions and share power.

hierarchy   Semi-autonomous levels that form from the interactions among a set of variables that share similar speeds (and geometric spatial attributes).

hysteresis   How a system responds, or more specifically, the return path taken following some disturbance or change due to cumulative effects. When the system follows a different path upon return to its former state, this is called a hysteresis effect.

panarchy   A model of linked, hierarchically arranged adaptive cycles that represents the cross-scale dynamic interactions among the levels of a system and considers the interplay between change and persistence.

political ecology   The study of the relationships and complex interactions between political, economic and social factors with environmental issues and changes.

poverty trap   A departure from an adaptive cycle; it results in an impoverished state, with low connectedness, low potential and low resilience.

regime   The set of system states within a stability landscape.

regime shift   Sudden shifts in ecosystems, whereby a threshold is passed and the core functions, structure and processes of the new regime are fundamentally different from the previous regime.

rigidity trap   A sustainable but maladaptive system, where potential is high, connectedness great, and resilience is high.

scale   The spatial and temporal frequency of a process or structure. Scale is a dynamic entity.

self-organisation   Self-organisation of ecological systems establishes the arena for evolutionary change. Self-organisation of human institutional patterns establishes the arena for future sustainable opportunity.

social capital   The aggregate of actual or potential resources that can be mobilised through social relationships and membership in social networks.

social ecological system (SES)   Integrated system of ecosystems and human society with reciprocal feedback and interdependence. Social ecological systems act as strongly coupled, complex and evolving integrated systems.

specified or specific resilience   The resilience 'of what, to what'; resilience of some particular part of a system, related to a particular control variable, to one or more identified kinds of shocks.

stability   The ability of a system to return to an equilibrium state after a temporary disturbance. The more rapidly it returns, and with the least fluctuation, the more stable it is.

stability domain   A basin of attraction of a system, in which the dimensions are defined by the set of controlling variables that have threshold levels (equivalent to a system regime).

stability landscape   The extent of the possible states of system space, defined by the set of control variables in which stability domains are embedded.

sustainability   The likelihood that an existing system of resource use will persist, without a decline in resource base or the social welfare it is able to deliver.

sustainable development   Development that meets the needs of current and future generations; the goal of sustainable development is to create and maintain prosperous and flourishing social, economic and ecological systems.

threshold   A level or amount of a controlling, often slowly changing variable in which a change occurs in a critical feedback causing the system to self-organise along a different trajectory.

transformability   The capacity to create a fundamentally new system when ecological, economic, or social (including political) conditions make the existing system untenable.

variables   Controlling variables determine the levels of other variables in the system (e.g. nutrient levels determine the density of algae in a lake). *Fast variables* are often those on which human use of systems is based, and fast social variables include management decisions and policies. *Slow variables* tend to change slowly – they include ecological variables such as population age structures or social factors such as culture or values.

Index

Page references in *italics* refer to figures and tables.